PRAIRIE

REVISED EDITION

PRAIRIE

A Natural History
of the Heart of
North America

CANDACE SAVAGE

Illustrations by Joan A. Williams

DAVID SUZUKI INSTITUTE

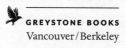

GREYSTONE BOOKS
Vancouver/Berkeley

Greystone Books Ltd.
greystonebooks.com

David Suzuki Institute
219 – 2211 West 4th Avenue
Vancouver, BC V6K 4S2

Cataloguing data available from Library and Archives Canada
ISBN 978-1-77164-594-2 (pbk.)
ISBN 978-1-77164-595-9 (epub)

Editing by Nancy Flight
Copy editing by Barbara Tomlin and Lara Kordic
Editing for revised edition by Paula Ayer
Proofreading by Alison Strobel
Cover and text design by Fiona Siu
Cover photograph by Drew Rush/Getty Images
Maps by the Canadian Plains Research Center

Printed and bound in Canada on ancient-forest-friendly paper by Friesens

Greystone Books gratefully acknowledges the Musqueam, Squamish,
and Tsleil-Waututh peoples on whose land our office is located.

Greystone Books thanks the Canada Council for the Arts, the British
Columbia Arts Council, the Province of British Columbia through the
Book Publishing Tax Credit, and the Government of Canada for support-
ing our publishing activities.

Canadä

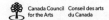

CONTENTS

PREFACE

It may be that when we no longer know what to do
we have come to our real work
and that when we no longer know which way to go
we have begun our real journey.

WENDELL BERRY, *STANDING BY WORDS*, 1983

DAY IN AND day out, we are confronted by news of impending disaster. • "Twelve Years to Limit Climate-Change Catastrophe" • "Warning of 'Ecological Armageddon' after Plunge in Insect Numbers" • "World's Food Supply under 'Severe Threat' from Biodiversity Loss." Each fresh pronouncement strikes a deeper tone of despair, until our ears are ringing with the clamor of doomsday bells. No wonder so many of us are stressed out and fearful. No wonder we often feel grief-stricken, angry, bewildered: overwhelmed.

Meanwhile, in my backyard in Saskatoon, Saskatchewan, on the northern edge of the great North American plains, three whorls of prairie crocuses, or pasque flowers, are pushing their fuzzy noses through the half-frozen earth for the first time since the prairie was torn up to make way for my house, ninety-odd years ago. The plants are there because two of my friends, Lisa and Renny Grilz, run a small business from their kitchen table, the Blazing Star Wildflower Seed Company. A few summers back, they collected crocus seeds along a soon-to-be-widened roadway, fingered them gently into pots—pointy ends barely covered, long plumes reaching up—and then tended the balky seedlings through two successive summers. I planted them out in my flower bed and crossed my fingers.

Last spring, our collective efforts were rewarded with a cluster of silky, mauve flowers and a harvest of feathery seeds; this year, the roots have sent up a dozen eager buds. Each of them offers a lesson in hope and resilience.

The prairies, the magnificent grasslands that sprawl across the heart of North America, have taken a ferocious beating in the last couple of hundred years, and the challenges that beset the Earth as a whole all echo loudly here. Climate chaos is already desynchronizing the emergence of butterflies and blossoms, and delivering alternate challenges of flooding and drought. Populations of grassland birds are in freefall, suffering the steepest and most persistent losses of any habitat group on the continent, decreasing by almost 60 percent in the last fifty years. Ten-thousand-year-old prairie continues to be torn up, replaced by cropland, coal mines, oil fields, and urban development.

Against this darkness, I offer you three crocus plants in a backyard. And I agree: as a response to environmental calamity, this is a woeful response. But what if those hopeful buds are just a beginning? What if, around them, we were to set the garden ablaze with prairie coneflowers, three-flowered avens, wild bergamot, giant hyssop, and buffalo beans? Let's send unruly outbursts of cleome, or stinking clover, through the gate and onto the boulevard, where they'll meet and mingle with a buzz-fest of bumblebees. Just for fun, we'll add neat labels beside the plantings out front, where the mangy lawn used to be, so that passersby will be inspired to begin restoring the prairie's wild beauty around their own dwellings.

And through this act of care and community, we may find that our minds begin to wander beyond their usual limits to the special places, near and far, where natural grasslands still survive, comparatively unharmed. Places where a shining wind ripples through stands of little bluestem and porcupine grass. Where the rising calls of curlews slice cleanly through the air. Where crocus buds push through the icy ground without aid from human hands. Merely to think of these light-filled spaces lifts our spirits and calms our thoughts. Despair is a useless emotion, and panic (though understandable) cannot help us find our way. It is Out There, in what Wendell Berry once described as "the peace of wild things," that we will find the courage to take the steps, large and small, that we need to be taking.

Here's what we already know: • We need to protect the remaining wild prairie and its unique animals and plants. • We need to restore the tilth and depth of prairie soils, not only to nourish ourselves but also to modulate the worst extremes of a changing climate. • We need to salvage, plant, restore, connect, lobby, and speak up.

At the same time, we need to make time for the pleasure of nurturing our own wild selves, by cultivating hope and renewing our capacity for joy and wonder. And that's where this book comes in. *Prairie: A Natural History of the Heart of*

North America provides a comprehensive overview of grass-land ecology and conservation, with a solid grounding in up-to-the-minute science. But this book is also a celebration, an extended love song to one of the world's largest and most remarkable ecosystems. Often overlooked and under-appreciated, the Great Plains Grasslands are an unbounded landscape of surprise and mystery. It's time to go exploring!

CHAPTER 1

WHERE IS HERE?

Prairies are part of the original fabric of the world.
CAROL DAVIT, MISSOURI PRAIRIE FOUNDATION,
AT THE NORTH AMERICAN PRAIRIE CONFERENCE,
HOUSTON, TEXAS, 2019

THERE ARE PEOPLE who think the prairie is boring, and it is hard not to pity them. We see them on the highways, trapped inside their cars, propelled by a burning desire to be somewhere else. But even as we wonder at their hurry, we have to admit that these disgruntled travelers are following in a grand old North American tradition. On both sides of the Canada–U.S. border, prairie bashing is as old as the written record. In 1803, for example, when the United States was contemplating the acquisition of the lands west of the Mississippi River, the great orator Daniel Webster was moved to object. "What do we want with this vast, worthless area," he thundered, "this region of savages and wild beasts, of deserts of shifting sands and whirlwinds of dust, of cactus and prairie dogs?" And even after this supposedly howling wilderness had been annexed to the United States, many observers remained unimpressed. The painter and naturalist John James Audubon was among them. In 1843, we find him traveling up the Missouri River on his first visit to the Great Plains. Forced onto the shore when his steamboat became grounded on a sandbar, he turned a disparaging eye toward the Dakota countryside. "The prairies around us are the most arid and dismal you can conceive of," he wrote. "In fact these prairies (so called) look more like great deserts."

Another traveler of the same era, a trader named Rufus Sage, was even more direct: "That this section of the country should ever become inhabited by civilized man except in the vicinity of large water courses, is an idea too preposterous to be entertained for a single moment." North of the border, Captain John Palliser, who crossed the Saskatchewan prairies in the late 1850s, was of much the same mind. Forget farming, he recommended. This country is just too dry.

It wasn't until near the end of the nineteenth century that the tide of expert opinion turned, and the Great Plains were opened to agricultural settlement, now touted far and wide as the new Garden of Eden. The fact was, however, that these magnificent grasslands were neither desert nor garden but something completely new to European and Euro-American experience. So new that at first there wasn't even a name for them in either French or English. Pressed to come up with something, the early French fur traders had extended their term for a woodland meadow—*une prairie*—as a kind of metaphor for this big, wide, sparsely wooded, windswept world. But the Great Plains were far more than a meadow. What the travelers had encountered was a vast, dynamic ecosystem, a kind of tawny, slowly evolving organism that, in a climate of constant change, had sustained itself ever since the retreat of the glaciers at the end of the Ice Age. In the presence of this strangeness and grandeur, words and vision failed.

When the newcomers looked around them, all they could see was where they weren't. This was not forest or sea coast or mountains; it was nothing but light and grass, the Big Empty in the middle of the continent. A vacant space, as they saw it, in desperate need of improvement. And this failure of vision—this inability to see and appreciate the Great Plains Grasslands for what they truly are—has continued to plague our perceptions right down to the present. Flat? Boring? Lifeless? Nothing could be further from the truth. It's time to drop out of the fast lane and give the prairies, our prairies, a second, loving look.

AN EMPIRE OF GRASS

THE KEY TO everything that happens on the prairies lies trampled under our feet. Although grasses may look humble, they are actually versatile and tough, capable of growing under the widest possible range of conditions. Anywhere plants can grow, grasses are likely to be on the scene, whether coexisting with cactuses in a desert, poking up among lichens on the Arctic tundra, or hiding in the leafy understory of a forest. And when circumstances are especially favorable for them—when the climate strikes just the right balance between precipitation and drought—grasses can assert themselves to become the dominant vegetation. ("Dominance," in this case, refers to the plants that contribute the most living tissue, or biomass, to the ecosystem. As trees to forest, so grasses to grasslands.)

A glance at a map of the world's major grasslands demonstrates that these conditions are most likely to occur on a broad, landlocked plain, far from any significant body of water, somewhere near the center of a continental landmass. It is in this semiarid environment—too wet to be a desert and too dry for forest—that grasses gain the upper hand, whether in the steppes of central Asia, the Pampas of Argentina, the savannas of Africa, or the broad heartland of North America. *See Map 16: The Great Plains; Map 1: Temperate Grasslands of the World.*

Globally, grasslands are among the largest of the Earth's terrestrial biomes, or life zones, with an expanse that covers more than a third of the land area of the planet. (At least, that's the area over which grasses would hold sway if natural conditions were allowed to prevail.) We're talking some 18 million square miles (46 million square kilometers)— almost three times the area of Russia. In North America alone, grasslands naturally extend over about 1.1 million square miles (2.9 million square kilometers), an area larger than many of the world's major nations.

The first European known to have set foot on this great domain of grass was a soldier and sometime explorer named Francisco Vásquez de Coronado. Dispatched from Mexico City in 1540, he was supposed to investigate rumors about a kingdom called Cibola, somewhere to the north, and to plunder its Seven Cities of Gold. When these glittering mirages turned out to be sunbaked Zuni pueblos in what is now New Mexico, he turned his attention to the uncharted Great Plains, where the fish were as big as horses, the people ate off golden

plates, and the king was lulled to sleep at night by a tree full
of golden bells. At least that's what people told him and what
he chose to believe. And so off set Coronado, with a party of
armed men, in the vague direction of present-day Kansas. In
the end the promised golden city turned out to be a town of
some ten thousand people who lived in grass-thatched homes
and sustained themselves by hunting bison and growing gardens, each in their season.

At every step of their journey, Coronado and his party
were astonished by what they encountered along their route.
Here lay "a wilderness in which nothing grew, except for
very small plants," but which nonetheless was teeming with
millions upon millions of strange humpbacked cattle. "I
found such a quantity of cows [bison]," Coronado reported,
"that it is impossible to number them, for while I was jour-
neying through these plains, until I returned to where I first
found them, there was not a day that I lost sight of them." Following these apparently endless herds were parties of hunting
people who dressed in bison-skin clothing (sewn with bison
sinew, drawn through a bison-bone awl), slept in bison-hide
tipis, and subsisted on a diet of bison blood and bison muscle.
Even the grass in this new world was cause for amazement,
as it rebounded from the conquistadors' steps and erased
the trace of their presence. In this great round world, all that
glittered was grass and an ecosystem of such richness and
diversity that it could scarcely be credited.

But think how amazed Coronado would have been if he
had somehow been able to sense the true extent and variety of North America's grasslands. Little did he know that
he had set foot on a vast prairie heartland—a continent of
grass—that was flanked on every side by smaller islands of

BUFFALO DANCE SONGS

Buffalo were not only a life-giving material resource to the Indigenous peoples of the Great Plains. They were also a spiritual presence. These songs were recorded around 1920 by a Pawnee man named Wicita Blain. Both had been passed down from previous generations and were inspired by dreams. The "waves of dust" in the first song seemed at first to hide a crowd of people, but the figures were soon revealed to be a herd of buffalo.

The Waves of Dust
Listen, he said,
There the buffalo are coming in a great herd,
These are his sayings.
There the buffalo are coming in a great herd,
The waves of dust roll downward.
There the buffalo are coming in a great herd,
They mark the place of the buffalo wallow.

The Buffalo Are Coming
Listen, he said, yonder the buffalo are coming,
These are his sayings, yonder the buffalo are coming,
They walk, they stand, they are coming,
Yonder the buffalo are coming.

Plains bison, Bison bison

grassland and prairie-to-forest transitions, or savannas. To the north, beyond his farthest imaginings, lay the Peace River Parkland, a region of rolling grass and poplars that marked the frontier between the Great Plains Grasslands and the boreal forest. To the east, the Prairie-and-Oak Transition zone—a tongue of prairie interspersed with groves of hardwoods—extended to the Great Lakes and beyond, marking the interface between the grasslands and the eastern deciduous forest. To the south, the prairies merged and melted into sultry, soupy marshlands to produce the semitropical vistas of the Western Gulf Coastal Grasslands. And to the west, in the broad valleys of the western Cordillera, lay the California Grasslands—spangled in spring by lupines and yellow-orange poppies—and the arid Palouse Grasslands of the Great Basin. Dominated by scraggly stands of sagebrush and spiky, sparse grasses, the Palouse, or bunchgrass, prairie stretched along the drainage of the Columbia and Snake rivers to intergrade with the shrubby growth of the Montana Valley Grasslands. *See Map 2: Temperate Grasslands and Savannas of Canada and the United States.*

And in the center of everything there lay the main attraction, the Great Plains Grasslands themselves, a landscape that even today invites wonderment. This truly is big sky country, with horizons that extend from the boreal forests of Alberta, Saskatchewan, and Manitoba to the deserts of the American Southwest and from the foothills of the Rockies to the Mississippi drainage. The numbers speak for themselves. Length: 1,500 miles (2,400 kilometers). Width: between 400 and 700 miles (between 600 and 1,100 kilometers). Vaguely triangular in outline, the region is broadest toward the north and narrows to its apex in the Hill Country of central Texas.

Total area: 1 million square miles (2.6 million square kilometers), or roughly 14 percent of the entire landmass of Canada, Alaska, and the Lower Forty-Eight States.

THE GRAND GEOGRAPHICAL TOUR

BUT LENGTH AND breadth are not the only descriptors of the Great Plains. The prairies also have a vertical rise and run that add a whole other dimension of interest. Formed primarily by sediments that washed out of the Rocky Mountains millions of years ago, the landscape slopes away from west to east, stepping down from an elevation of about 1 mile (roughly 1,600 meters) above sea level at the base of the foothills to a few hundred yards above sea level on the banks of the lower Missouri River. Often, the change happens so gently that you hardly notice it. Who would have imagined, for example, that the drive across Kansas, from west to east, following in Coronado's path, would be downhill all the way and that you'd lose more than half a mile (a kilometer) in elevation while traversing that seemingly level state?

Overlain on this gently sloping plain are a surprising diversity of landforms. The geography of the Great Plains offers something for every taste, from fantastically sculpted badlands to craggy mountains to some of the flattest expanses of country anywhere on the planet. "I reached some plains so vast, that I did not find their limit anywhere I went," our old friend Coronado exclaimed in a letter to the king of Spain in 1541, "with no more land marks than if we had been swallowed up by the sea... not a stone, nor bit of rising ground, nor a tree, nor a shrub, nor anything to go by." The landscape to which he was referring is now known to geographers as the High Plains, an elevated and sometimes spectacularly featureless tableland that extends from Nebraska and Colorado into northern Oklahoma and Texas. An erosional remnant of a high-and-wide landscape that once extended over much of the Great Plains, the region is bounded on three sides by dramatic cliffs, including the upthrusting wall of the Mescalero Escarpment in the west, the tree-clad Pine Ridge Escarpment to the north, and the amazingly convoluted and striated Caprock Escarpment in the east. *See Map 3: Geography of the Great Plains.*

To the south of the High Plains lie the limestone hills of the Edwards Plateau, or Texas Hill Country—a world in itself—where the rolling countryside is broken by domed upwellings of rock, deeply cut by streams, and eaten away underground to form a honeycomb of sinkholes and caves. The Edwards Plateau, in turn, is bounded on the south by the terraced ridges and eroded canyons of the Balcones Escarpment, which slashes across Texas at the southern limits of the Great Plains Grasslands.

To the northwest of the Edwards Plateau lies the broad Pecos Valley and a landscape of spectacularly eroded caverns, sinkholes, and steep-walled limestone cuts. And north of the Pecos are the shadowed moonscapes of the Raton Section, where mesas capped with lava compete for attention with contorted badlands and the burned-out cones of Capulin Mountain and other long-extinguished volcanoes. From there, it is on to the broad, terraced valleys of the Colorado Piedmont, literally "foot of the mountains," where the waters of the Arkansas and South Platte rivers have, over millions of years, stripped away layer after layer from the original High

Plains surface. (This dramatic, if localized, lowering of the surface explains, for example, why the road heading east out of Denver tracks steadily upward for the first half hour or so, as it climbs out of the South Platte floodplain and onto the surrounding High Plains benches.) The effects of water erosion can also be seen on the rugged Missouri Plateau and the deeply dissected valleys of the Plains Border region.

If water has cut into these landscapes, wind has smoothed them out. For example, the southeastern edge of the Platte River valley is softened by a broad belt of curving, undulating sand dunes that were deposited by dust storms sometime during the Ice Age. Similar formations, shaped by similar forces, are also to be found strewn up and down the drier, western side of the Great Plains, from the Great Sandhills of southwestern Saskatchewan in the north to the Mescalero Sands of the Pecos Valley. And right in the middle of the map lies one of the prairies' little-known natural wonders—the Nebraska Sandhills, a region of whale-backed, grassy rises and prairie wetlands that, at an area of almost 24,000 square miles (61,000 square kilometers), ranks as the largest field of sand dunes in the Western Hemisphere. These sandscapes were put in place by the relentless northwest winds that have been coursing across the landscape for millions of years.

With so few barriers to stand in their way, these same winds have had the run of the entire Great Plains region. Although their influence can be seen in many parts of the country—for example, as ridges of windblown silt along both the South Saskatchewan and Upper Missouri rivers—their influence is most obvious in the eastern and southern regions of the Great Plains. These include not only areas of the Colorado Piedmont and the High Plains but also the "low plains" to the east, notably the rolling hills of the Plains Border country, the Osage Plains, and the Glaciated Central Lowlands. Much of this sweep of country is blanketed in deep, contoured drifts of fine silt, or loess—pronounced "luss"—another gritty, wind-borne by-product of glaciation. The result is a gently undulating landscape of soft, rolling hills and, in places, extraordinary bluffs, like the delightfully eroded and unexpected Loess Hills of western Iowa.

The northern plains region, by contrast—north and east of the Missouri River, from Alberta to Manitoba and south through the Dakotas—is less apt to be buried in loess, but it nonetheless bears the imprint of the Ice Age. Here the terrain is an unmade bed of glacial rubble, or till, lying exactly where it dropped when the ice sheets retreated from the landscape thousands of years ago. And protruding above this jumble of knobs and kettles is an assortment of sprawling, flat-topped uplands, including Turtle Mountain, Wood Mountain, and the Cypress Hills, which straddle the boundary between past and present. Like miniature versions of the High Plains, they are the last surviving remnants of an ancient, preglacial landscape that has otherwise been lost to erosion.

Finally, and most surprising of all, are the honest-to-goodness mountains that jut out of the northern plains, particularly on the unglaciated reach of country south and east of the Missouri River. From the glowering Black Hills to the jagged Crazy Mountains, they stand as a peak experience (if you'll forgive the pun) for anyone who has been led to believe that the prairies are monotonous.

THEN AND NOW

IT IS ONE thing to send our minds running across the contours of the Great Plains Grasslands and their unexpectedly varied landforms. It is quite another to bring these spaces to life, to try to perceive them in their full, natural vitality and splendor. What would it have been like to step out onto the round bowl of the southern grasslands with Coronado in 1541, aware that at any moment our progress might be blocked by a dusty, pawing, milling herd of bison? Or, precisely 150 years later, in 1691, to have traveled with Henry Kelsey and his Nēhiyaw and Nakota guides from Hudson Bay through the northern forest and onto the prairies of the Saskatchewan River country? What emotion would have seized us when a blocky, hunched shadow gradually resolved into the form of a massive and potentially lethal grizzly bear? Or what if we could slip back in time to 1804–6 (little more than two hundred years ago) and join Lewis and Clark on their famous expedition up the Missouri River?

Plains grizzly bear

Imagine: Bison beyond counting. ("I do not think I exaggerate," Lewis wrote as he crossed the Dakota plains in 1804, "when I estimate the number of Buffaloe which could be compre[hend]ed at one view to amount to 3000.") Flights of pronghorns at every turn. Elk coming up out of misty valleys to graze on the prairie at dawn. Bighorn sheep perched on the steep, crumbling walls of the Little Missouri badlands. Wolves threading across the prairies, trailing the herds.

Two hundred years isn't very long on the geologic timescales of planet Earth. These memories lie at the very threshold of the present, so close that we half expect to be able to walk into a fold in the landscape and encounter them. And something like this still occasionally happens when we stumble across a physical trace of the past, whether it's a flaked stone tool that once belonged to a bison hunter or a shallow, saucer-shaped hollow that was worn into the dirt by generations of rolling, grunting bison. The animals have vanished, but the imprint of their flesh and blood is still on the land. It is all so mind-bogglingly recent.

There are not many places where the wild is as close at hand as it is on the Great Plains. In Europe and Asia, no one can quite remember what "natural" looked like, because the land has been successively shaped and reshaped to meet human needs for hundreds or thousands of years. But on the prairies—right up to the moment when agricultural settlement began—humans had lived off the natural productivity of this vast, sun-swept expanse of grass. The First Peoples drew their sustenance from the animals and plants, experiencing both feast and famine as hunters and gatherers. This is not to say that they sat back passively and let nature take its course. They were active participants in the ecosystem, ready and willing to use whatever technologies they could command to improve their chances of survival. For example, they had no qualms about setting the prairies on fire, to green up the grass and draw bison in for the hunt. They tilled the soil of fertile river valleys and planted gardens of sunflowers, corn, and squash. They eagerly adapted to the new culture of firearms and horses.

Yet despite these human innovations, the underlying dynamic of the ecosystem—the interplay between climate and grasses, grazers and predators—remained robust. A landscape that had evolved to support large herds of grazing animals was still doing exactly that, as life ebbed and flowed in time with the seasons. Then, in the early to mid-1800s, the pace of change accelerated. In far-off Washington and Ottawa, ambitious governments began to assert their claim to the land and resources of the Great Plains. As a prelude to agricultural settlement, Indigenous people were confined on reserves and reservations, whether by persuasion or by brute force, and the bison on which they depended—the multitudes of "humpbacked cattle" that had darkened the plains—were virtually wiped out in a bloody orgy of killing. Tellingly, the final stages of this slaughter were motivated by the discovery that bison hides could be cut and sewn into leather belts and used to power machines in the burgeoning industrial complex in the East. (The last free-roaming bison were killed in Canada in 1883 and in the United States in 1891.) Modern times had arrived on the prairies.

And then came the settlers, an onrush of humanity that reached full flood in the late 1800s and early 1900s. Determined to make a stand in this challenging country, the incomers quickly progressed from temporary shacks and shanties into substantial homes, making them the first people ever to establish permanent, year-round dwellings on the open plains. This was a bold experiment, occasioned with far more risk than anyone at the time seemed to recognize or, at least, was prepared to admit. But whatever the hazards, the way forward was clear. The object was to assert control over the ecosystem and redirect its natural vitality into the production of commodities that could be bought and sold on the world market. Beef, not bison. Wheat and corn instead of prairie wool.

The result of this revolution is the landscape that we see today, a colorful patchwork of fields and rangelands, where geese feed in the stubble, foxes hunt in farmyards, and meadowlarks sing their hearts out on fence posts. These are the prairies that our generation was born to, and they are beautiful in their own right.

Yet the more we love this place as it is, the more we feel the pain of what it so recently was. The wild prairie ecosystem is gone. And this tragedy is compounded by the realization that we don't even know exactly what it is that we have lost. "Civilization" and "progress" overran the grasslands with such an urgent rush that the ecosystem was disrupted before anyone had a chance to make a systematic study of exactly what was here or to figure out how all the pieces interacted with each other. The people who might have had the most to teach us—the last generation of hunters and gatherers—went to their graves largely unheeded by the newcomers, taking their knowledge of the prairie and its life ways with them. We are left with little to guide us except for knowledge handed down by Elders and fragments of written descriptions in the journals of explorers and early settlers—partial lists of species, brief sightings, and offhand remarks—that leave many basic questions unanswered.

The depth of our ignorance is startling. Question: How many bison were there on the plains before the slaughter began? Answer: No one can tell us with any assurance. By working and reworking the available strands of evidence, experts have estimated the precontact population at anywhere from 12 million to 125 million animals, a variance that leaves more than 100 million bison in limbo. These days, experts acknowledge that bison once numbered in the millions and probably tens of millions, but that's as far as they're prepared to go. And if we cannot account for big things like bison, how much less do we know about the smaller and less conspicuous organisms—little things like insects and spiders, fish and frogs, rodents and songbirds—that lived and died in their untold variety and interest and abundance? Yet if the wild past is lost to us, we can still look ahead. Despite everything that has happened, it is not too late to acknowledge the natural forces that continue to animate the prairie world and that, even today, shape the lives of all its creatures.

ECOREGIONS OF THE GREAT PLAINS

"ECOLOGICAL LAND CLASSIFICATION" is an attempt to see beyond the human impacts of the last few centuries and uncover the enduring components of the environment (climate, soils, landforms, vegetation, and so on) that make one part of the world biologically different from the next. While we cannot go back in time and view the wild prairies in full bloom, we still have a chance to identify and assess the fundamental factors that define them.

Over the last decade, the ecological regions of the Great Plains have been mapped in different ways by different agencies, whether in broad strokes as part of continentwide research or more minutely, state by state and province by province. One result of this effort is a set of maps created under the joint authorship of the World Wildlife Fund Canada and its counterpart in the United States, the wwf–U.S. Seen through the lens of these organizations, the Great Plains Grasslands come into focus as a mosaic of fifteen ecoregions. *See Map 4: Ecoregions of the Great Plains.*

For example, the Aspen Parkland ecoregion lies across the midriff of the Prairie provinces (Alberta, Saskatchewan, and Manitoba) and provides an interface between the boreal forest and the open plains. Rising in the north as a closed poplar woodland with occasional stands of spruce, it gradually unfolds into a rolling grassland dotted with aspen groves and dominated by various speargrasses, wheatgrasses, and most notably, fescues.

Rough fescue

Because of the predominance of fescues in the Aspen Parkland, the region is sometimes known as a fescue grassland. The same term is also applied, for the same reason, to the community of plants found in the Foothills Grasslands. Located on the undulating slopes at the foot of the Rocky Mountains, this ecoregion is dominated by rough fescue, together with lesser quantities of June grass, speargrass, wheatgrass, and various flowers and shrubs.

East of the foothills and south of the Aspen Parkland lie two ecoregions with subtly different characteristics. The Northern Mixed Grasslands take their name both from their northerly location and from their characteristic mixed-height cover of grasses. Here, the sparse, tufted vegetation of the foothills gives way to a groundcover of ankle-high grasses, notably blue grama, intermixed with an overarching canopy of knee-high stands, including various spear- and wheatgrasses. As this mixed grassland flows southward, the canopy of taller, midheight grasses gradually thins out, and the overall height of the vegetation diminishes. This transition from mixed- to short-grass prairie continues to the south in the Northwestern Short/Mixed Grasslands ecoregion.

The trend to shorter grasses culminates in the Southern Short Grasslands of the High Plains. Here, in an area once memorialized by Coronado as a land of "very small plants," the vegetation is dominated by a ground-hugging mat of grama and buffalo grass. Yet here, too, there are subtle transitions. For if the grasslands diminish in height from north to south, they shoot up from west to east as they escape from the rain shadow of the Rockies. This trend is reflected in the shift from the short-grass prairies of the west, with their carpet of stunted plants, to the multilayered, knee-high vegetation of the Nebraska Sandhills and the Southern Mixed Grasslands.

On the eastern flank of the plains lies the tall-grass-prairie region, so named for the luxuriant stands of big bluestem and Indian grass that once flourished there. "The grass is so very high that a man is lost amongst it," one early traveler reported from southern Wisconsin in 1761. Bright with brown-eyed susans and other flowers, these magnificent prairies extended from the Northern and Central Tall Grasslands south through the Flint Hills to the Blackland Prairies of east-central Texas. Many of the same species of grasses are also found, as an understory, in the juniper breaks of the Edwards Plateau Savannas and in the hickory-and-oak woodlands of the Cross Timbers Forest and the Southern Prairie-and-Oak Transition.

And finally, right out in the middle of everything, stand the lonely, displaced islands of ponderosa pine, white spruce, and paper birch that make up the Black Hills Coniferous Forest.

The ecological interactions that find expression in these varying landscapes have been at work for thousands of years. Even today, characteristics such as average temperatures, precipitation, length of growing season, and drainage patterns provide the physical framework or, one could say, the loom on which the fabric of the Great Plains ecosystem is woven. Yet for all their continuing importance, these long-term physical features are no longer the only powers in the land. Other interests have taken over; other hands are pulling threads. Those hands, of course, are human.

Over the last two hundred years, human beings have hit the prairies with the force of a major geological crisis, triggering not only extinctions and extirpations—of plains wolves, plains grizzlies, plains elk, plains bighorn sheep, free-ranging plains bison—but also dramatic shifts in the vegetation. Taken as a whole, the Great Plains Grasslands now rank as one of the most extensively altered ecosystems on Earth. There is scarcely a patch of ground where we have not left our footprints. The southernmost Short Grasslands, for instance, are as sun-frazzled and arid a country as you could ever expect to see, yet even there an estimated 27 percent of the ecoregion has been brought under cultivation. The surviving native prairie in the region is now devoted to livestock or converted to ranchettes, on the advancing front of urbanization. In the mixed grasslands, by contrast, the percentage of land under cultivation rises from 15 percent (in districts with scant precipitation) to over 96 percent (where conditions are most conducive to crop production). And in the tall grasslands, with their relatively generous climate and deep, black earth, fully 99 percent of the native grasses have been plowed under to make way for farming.

Largely as a result of this destruction of natural habitat, at least 464 prairie species have declined to such rarity that their long-term survival is in question, and more names are added to the list with every passing year. (This tabulation includes only species that have been officially designated as at risk of extinction, either locally or nationally.) Of this total, a majority are unique organisms found exclusively, or almost exclusively, on the Great Plains Grasslands. Last chance to see the Dakota skipper butterfly, the Colorado checkered whiptail lizard, Preble's meadow jumping mouse, the black-footed ferret.

These trends are deeply troubling, and we could easily get lost in the dark. To find our way forward we will have to be sure-footed, willing and able to move quickly from sorrow to hope, from past to present, from celebrating wildness to accepting and honoring our own accident-prone presence. We will need to see both the splendor of the life that has faded away and the abundance that still extends across the whole

Chimney Coulee, June 29: "Last night, we stood on a hillside, ankle-deep in prairie wool, and heard a quavering whisper that seemed to come out of nowhere. And again, like a sudden sigh. Finally we saw them, high above our heads, a pair of nighthawks that sometimes interrupted their insect-hunting maneuvers to plunge headlong down the sky and rasp the air with their wing feathers. In that moment, the whole place was shot with silver."

Grasslands National Park, July 9: "In just two days we have seen meadowlarks, horned larks, Sprague's pipits with their surprising pink feet, phalaropes spinning and dipping like wind-up birds in a dugout, sharp-tailed grouse, golden eagles (nestlings and mature), jackrabbits—huge, hallucinatory—cottontails crouching in the shade of boulders in a prairie dog colony, mule deer, white-tails, pronghorns (bucks, cows with calves), monarch butterflies, showy milkweed in full bloom, prickly pear cactus with waxy yellow flowers, jumping cactus stuck to our dogs, purple prairie clover, silverleaf psoralea, brown-eyed susans, skeleton weed, blue Missouri milkvetch against bone-white clay, needle-and-thread grass, awned wheatgrass shining in the wind. The cold white glare of a full moon."

Aspen Cut, November 11: "A bright blue sky, fresh snow, sparkling and mild. We are standing at the edge of a wooded draw when my companion says, 'A coyote!' and points to the opposite slope. But it's not a coyote. It's a cougar: reddish-brown, stocky, long heavy tail. It flows up the slope—looks back over its shoulder—then over the ridge and gone. On a snow-covered log, we find large round tracks with pinprick claw marks."

wide world of the prairie in every direction. For however diminished, the Great Plains are blooming and buzzing and wriggling and squirming with wildlife wherever we look. In the Northern Mixed Grasslands ecoregion, for example— where as much as three-quarters of the natural habitat has been lost to the plow—there are currently no fewer than 13 species of amphibians, 18 reptiles, 72 mammals, at least 160 butterflies, 222 birds, and 1,595 species of grasses, sedges, and wildflowers. This gives the region a total "species richness index," on the books of the World Wildlife Fund, of 2,095, much higher than many areas that are typically thought of as biodiversity hot spots. (By comparison, the rain forests of northern California have a richness index of only 1,710, while the Everglades come in at 1,855.)

On the Southern Short Grasslands, where significant areas of natural grasslands remain intact as grazing land, the picture is brighter yet, with 17 species of amphibians, 61 reptiles, 86 mammals, 230 butterflies, 245 birds, and an astonishing 2,359 species of grasses and other nonwoody plants, for a richness index of 3,011. Although this book can't introduce you to all those species—you'll need the appropriate local field guides for that—it will explain how this abundance of life is sustained and renewed, season after season. Far from being a sacrifice on the altar of progress that we can dismiss from our thoughts, the prairies are still very much alive and worth caring about.

ABOUT THIS BOOK

IN THE RIOTOUS interactions of nature, everything happens at once—sun, wind, rain, growth, birth, death—and change ripples organically through the ecosystem. For the purposes of discussion, however, it has been necessary to isolate aspects of this holistic system and discuss them one by one, each in a separate chapter. Although the subject matter is tightly interrelated, each section has been designed to stand on its own, so the chapters can be read individually and in any order. Chapter 3, "The Geography of Grass," for instance, provides a detailed look at the prairie grasses and their dynamic relationship with the extremes of a midcontinental climate. Chapter 4, "Secrets of the Soil," ventures into the dirt—a life zone all its own—and introduces a few of the strange little creatures that live beneath the ground. In Chapter 5, "Home on the Range," we come back out into the sunshine to ride through cattle country and find out how life is lived on the surviving expanses of native prairie. Chapter 6, "Water of Life," by contrast, takes us knee-deep into the nearest prairie river or pond to look into the lives of ducks, shorebirds, fish, and other aquatic organisms. Chapter 7, "Prairie Woodlands," examines the unexpectedly important role of trees in grassland ecology and asks what difference it makes that woody growth is now invading the prairies. Chapter 8, "The Nature of Farming," studies the potential of croplands to provide wildlife habitat. And finally, Chapter 9, "Long-Range Forecast," reconsiders the conservation status of the Great Plains—is this really the most endangered ecosystem on the continent?—and discusses a range of options for protecting and restoring its wildness.

But before we look to the future, Chapter 2, "Digging into the Past," will take us back to the very beginnings of time and the great adventure of life on Earth.

CHAPTER 2

DIGGING INTO THE PAST

A prairie requests the favor of your closer attention.
It does not divulge itself to mere passersby.

SUZANNE WINCKLER, *PRAIRIE:*
A NORTH AMERICAN GUIDE, 2004

THERE IS AN unseen dimension to the prairies, and that dimension is time. At first glance, this landscape may seem to have avoided the ravages of the past. The level plains and soft, rolling hills appear to have settled here quietly, their surface unmarred by signs of geological strife. But appearances can be deceiving. The great grasslands of central North America have been shaped over the past three or four billion years by the same forces that raised the Rockies and excavated the Grand Canyon. Their surface has been seared by the sun, scoured by ice, blasted by blowing sand, and buried in deep drifts of gravel. As a result of immense energies beneath the surface of the Earth, the plains have been raised up, forced down, drowned by oceans, and blanketed in ash.

They have experienced every shudder and wrench as continents have collided and torn away from each other, only to collide and tear away again.

The traces left on the surface of the prairies by this planetary bump and grind are surprisingly minimal. Yet if you know what to look for and where to look for it, the subtleties of the prairie landscape become eloquent. An oil well bears witness to ancient tropical seas. A vast level plain provides an unexpected reminder of the protracted violence of mountain building. A hummocky wheat field speaks of the lumbering passage of glaciers. To an observer with a little basic geological knowledge, even the most unspectacular prairie landscape suggests a long and spectacularly interesting history.

UNDER THE WAVES

TO GO BACK into the prairie's history means to go down. The record and residue of times past lie beneath our feet, so wherever we go on the prairies, we are traveling across vanished worlds. Straight beneath you, for example, at a depth of between 1,800 and 4,000 miles (2,900 and 6,400 kilometers), lies the Earth's core—the yolk of the planetary egg. This partly solid, partly fluid center is encased in an equally ancient layer of rock called the mantle. Surrounding the mantle, in turn, is a covering of waxlike malleable material known as the asthenosphere, which is kept at a lethargic boil by the heat of its own radioactive decay. As the source of the molten magma that periodically shoots up through volcanic fissures and rifts in the ocean floor, the asthenosphere is the main powerhouse of geological turmoil.

The asthenosphere is overlain by a relatively thin and fragile shell of rock known as the lithosphere. The outermost membrane of this rocky shell is the Earth's crust, a layer that is thinner, proportionately speaking, than the skin of an apple. On the prairies, the crust extends to an average depth of 25 to 30 miles (roughly 40 to 50 kilometers). Yet this comparatively short vertical distance takes us back in time some 3.8 billion years, to an era when the flying debris of creation had begun to subside and the Earth's crust was finally able to stabilize. In this remote and inhospitable age, we find the first traces of life—microscopic stains, a few microns long, made by filaments of cyanobacteria, or blue-green algae.

Rocks from this primordial era lie right out in the open on the Precambrian Shield, but they seldom break through to the surface of the Great Plains. Instead, these ancient granites generally lie several miles beneath our feet, providing the foundation, or "basement rock," on which the prairies have been built. Our region lies on what geologists call the North American craton, the stable core of the continent. This is a large fragment of the Earth's crust that sheared away from an unnamed supercontinent toward the end of the Precambrian Era. By the time this happened—some 600 million or 700 million years ago—the Earth (and the prairie region along with it) had already endured more than three billion years of mountain building, erosion, glaciation, deglaciation, and general geological *Sturm und Drang*. But things must have been starting to settle down, because when the

GEOLOGICAL TIMESCALE

AGE	ERA	PERIOD	EPOCH	YEARS AGO
	Cenozoic	Quaternary	Holocene	10,000
		Tertiary	Pleistocene	1.6 million
			Pliocene	5 million
Age of Mammals			Miocene	24 million
			Oligocene	34 million
			Eocene	55 million
			Paleocene	65 million
	Mesozoic	Cretaceous		145 million
Age of Reptiles		Jurassic		200 million
		Triassic		250 million
	Paleozoic	Permian		300 million
Age of Amphibians		Carboniferous		355 million
Age of Fishes		Devonian		420 million
		Silurian		440 million
		Ordovician		490 million
		Cambrian		545 million
	Precambrian			4.5 billion

supercontinent tore itself apart, it produced a North American continent-in-the-making that has persisted until the present.

This infant continent was not exactly the landmass that we know today. The entire Western Cordillera was missing, with the result that the west coast of the craton ran south through present-day British Columbia and the Pacific states (much closer to the prairies than it is today). At first, the cratonic landmass lay exposed—a low, eroding plain, as barren as the face of Mars. But, as the geological strife continued, sea levels began to rise and the land was gradually overrun by the ocean. In time, the entire continent (with the periodic exception of a chain of tropical islands that ran diagonally across the plains, from Lake Superior toward Arizona) had disappeared beneath the waves.

For roughly the next 55 million years (from about 545 million to 490 million years ago), much of the North American craton lay under a shallow sea. Wherever the land remained exposed, it was eroded by water and wind, which ground the gritty Precambrian rocks into rounded grains of quartz sand. This sand was then swept to the coasts and out into the sea, where it settled to the bottom in beds tens to thousands of yards (or meters) thick. Eventually, these lustrous sediments were overlain by layers of fine-grained mud. But whether sandy or silty, this ocean floor was literally crawling with life, particularly three-lobed, many-legged, bottom-feeding arthropods known as trilobites. After an agonizingly slow start with the cyanobacteria, evolution was finally hitting its stride, producing a menagerie of weird and wonderful undersea life. As generation upon generation of these animals lived and died, their remains settled onto the ocean floor, where they were buried under thick layers of sediments. Today these fossil-rich deposits—now compressed into solid sandstone and shale—are buried some 3 miles (5 kilometers) beneath the wheat fields of the northern plains and at lesser depths in other parts of the prairies. But in a few places—like the Judith and Little Rocky mountains and the northern Black Hills—they have been pushed up to the surface, exposing their maritime history to plain view.

When the Cambrian sea finally withdrew and dry land emerged again, the forces of erosion immediately began to tear away at the newly formed rocks. But soon, geologically speaking—after a break of little more than 20 million years—the water rose and slowly spread over the land. This time, even the transcontinental island chain was bathed in the warm, clear seas. Now primitive snails munched on algae and were themselves preyed upon by giant squidlike nautiloids, with shells up to a couple of yards in length. Hundreds of new species of shelled animals evolved, including crinoids, or "sea lilies" (distantly related to modern sea urchins), and exotic reef-forming corals. There was so much life in these oceans that when they finally withdrew some 440 million years ago, they left behind thick deposits of shell fragments and calcium-rich debris, which eventually solidified into fossil-rich limestones. These Late Ordovician deposits include the elegant Tyndall stone that is quarried in Manitoba and graces so many buildings in the Prairie provinces.

And so things continued for about the next 100 million years, as shallow oceans advanced across the North American craton, only to withdraw and then flood back in. If the run and roll of the grasslands sometimes make us think of the sea, surely this is a reminder of the landscape's long marine history. With every advance and retreat of the ocean, the land was burdened with fresh deposits of sand, silt, and crushed shells, which built up, year by year, in nearly horizontal, banded layers. Although some of these contrasting sea floors have since been exposed by erosion (where rivers have cut deeply down through the sediments), for the most part they lie thousands of yards (or meters) beneath the grasslands.

The oceans that left these deposits behind were hospitable to life—shallow, warm, well lit, and typically tropical. During the Silurian and Devonian periods in particular (between about 440 million and 355 million years ago), these waters provided ideal conditions for reef-forming sponges and corals. In what would one day become the Canadian Prairie provinces, the reef builders of the Devonian had a heyday, constructing barrier reefs and ringlike walls that rose to heights of 300 feet (90 meters). Wherever the sea was constricted by these limestone palisades, the water gradually became super-salty. If the circulation of the sea was inhibited, water lost to evaporation could not readily be replaced, and the concentration of salts steadily increased. In time, the salts precipitated out of the sea water in these areas, leaving thick beds of potash and other salts, notably under present-day Saskatchewan. The potash deposits in Texas were formed by a similar process but some millions of years later, during the Permian Period.

The last truly continentwide inundation withdrew from the North American craton about 300 million years ago. The next time the sea attempted to overrun the land—as it would continue to do for millions of years to come—it found itself lapping around the shores of a rocky upland that had started to rise in the eastern half of North America. Apparently, the asthenosphere had heated up and begun to force masses of molten rock up through rifts in the ocean floor. This event had sent the continental plates on a slow and perilous collision course. First, Europe smashed against North America from the northeast. Then a massive supercontinent called Gondwana (the combined landmasses of South America, Africa, India, Antarctica, and Australia) crunched into North America from the south, causing the land to buckle and forcing the Appalachian Mountains to lift along the east coast. The forces involved in these mighty adjustments were even felt in the middle of the craton, where a range of mountains 3,000 feet (900 meters) high rose out of the plains of present-day Oklahoma and Colorado. Known as the Ancient Rockies, these mountains have since been eroded to their roots by the action of water and wind.

Apart from the appearance of these new highlands, the west coast of the craton was comparatively untouched by these titanic collisions. Through all the commotion, the sea continued to wash up over the land, even splashing around the base of the Ancient Rockies and turning them into a cluster of islands. With every advance and retreat, the sea again left behind layers of sediment and the fossilized remains of a strange coterie of underwater life. In addition to the crinoids, corals, and other unusual beasts that had occupied Devonian waters, there were now small filter-feeders, called archimedes, that had perfect corkscrew skeletons, and others, called productids, that held themselves up off the sea bottom by perching on stiltlike spines. (A wonderful jumble of 250-million-year-old sea life has been preserved in the Guadalupe Mountains of western New Mexico and Texas, which were once a complex of reefs in the western ocean.) Bony fishes swam through these waters, sometimes hotly pursued by large, saw-toothed sharks. The game of evolution was being played with feverish exuberance.

Nautiloids   Productid

Meanwhile, back on dry land, the surface of the continent was continuing to buckle and twist. As the Appalachians were thrust upward, land in the center of the craton was forced to rise along with them. A broad plain formed along the edge of the eastern highlands, sloping gently toward the western sea. When the waters receded, this coastal plain extended all the way west to present-day Alberta and Texas. And even when the sea rose up and flowed across the land, the eastern margin of the plain (roughly from present-day Manitoba south to Kansas and Missouri) was now high enough to escape all but the most severe flooding.

A new frontier for life was emerging not only in North America but on the other continents as well. Land plants, which had put in their first appearance some millions of years before, had never made much of a showing. But as stable new habitats became available, the evolutionary tree began to bud and sprout with explosive energy, producing more and larger species of land plants than ever before. In time, the soggy, boggy landscapes left by the retreating oceans were filled with riotous jungles of giant sphenopsids, or scouring rushes, tree-sized ferns and leafy conifers. These tremendous swamps, which flourished between about 355 million and 300 million years ago, disappeared soon afterward, probably as a result of a cooling and drying trend in the climate. Buried where they fell—in modern-day Iowa, Missouri, and Kansas, among other places—the swamp plants eventually turned into coal, the characteristic rock of the Pennsylvanian, or Upper Carboniferous, Subperiod.

Through much of the next 50 million years (the Permian Period), the land shriveled in the sun. Swamps decayed, seas shrank, and the exposed plains along the west coast blew with sand and salt. But life was not to be stopped. Insects, which had dominated the wetlands of the Carboniferous, now gave rise to new dry-land forms such as beetles and the distant ancestors of crickets and grasshoppers. Amphibians, too, crawled out of the swamps and began to invent the technology they needed for life on the land—notably a soft-shelled, amniotic egg that could develop out of the water. In time,

Dimetrodon

new life-forms developed that could live their whole lives on land, including massive, lizardlike creatures known as stem reptiles. Basking alongside these primitive organisms on the arid coastal plains were their near relatives, the synapsids—the direct ancestors of modern mammals. At the root of our family tree is *Dimetrodon,* a burly, fin-backed synapsid with two stabbing canine teeth, which it used to snap up slow and unwary amphibians. We know these creatures once roamed the savannas of the western plains because wonderfully preserved skeletons of *Dimetrodon* and many of its equally bizarre contemporaries have been dug out of Permian "red bed" deposits in New Mexico, Texas, and Oklahoma.

And so it is that we find ourselves near the end of the Permian, watching a lumpish, beaked synapsid called *Kannemeyeria* breaking off the tough stem of a broad-leafed conifer somewhere along the west coast of Texas. Under our feet lie the accumulated sediments of 3.5 billion years, or more than 90 percent of the geological timeline. Yet except for the wide spread of the horizon, there is little in this scene to put us in mind of the modern prairies. No grass, no gophers, no pronghorns, no playas or sloughs. Something radical will have to happen to create the landscape that we see around us today. Something revolutionary.

TERRIBLE LIZARDS

THE PERMIAN PERIOD ended in a biological catastrophe—the most severe mass extinction in all of geological history. During a period of several million years, over 95 percent of all the species living in the oceans were eliminated, together with 75 percent of terrestrial vertebrates. Why did this disaster occur? No one knows for sure, but the continuing gyrations of the continental plates may have been partly to blame. By the Permian Period, the continents had become temporarily fused into one gigantic landmass, called Pangaea. At the same time, the floor of the oceans apparently warped downward, drawing the sea away from the land and exposing a vast and inhospitable heartland of hot, dry silt and sand. These deserts had little to offer to life-forms that, in ages past, had flourished in a watery world of lagoons and swamps. Perhaps this change in conditions is enough to explain the huge loss of life. But whatever the probable causes (and many have been invoked), the impact was severe, and, despite the non-stop creativity of evolution, it would take millions of years for the Earth to repopulate itself with a full range of plants and animals.

At the same time that this biological revolution was occurring, a major geological upheaval was also underway. The continents, after docking together in Pangaea for some millions of years, began to tear away from one another. As Europe sheared off to the east and the Atlantic Ocean opened up, the North American craton was shoved slowly westward. Eventually, about 165 million years ago, the drifting continent ran into a small fragment of the Earth's crust (perhaps an island chain), known to geologists as a terrane. As the continent plowed onward, it contacted other, similar obstacles in its path. One by one, these terranes were crushed against the west coast of the craton and added to its mass. The impact of these collisions—which would continue sporadically for about the next 100 million years—caused the western margin of the craton to fold, twist, crack, and rise up mightily, until

THE EARTH IN UNSTEADY MOTION

Sometime in the first half of the nineteenth century, a man named Little Hill, "a Winnebago chief from the upper Mississippi, west," shared this story with a U.S. government official and amateur ethnographer named Henry Schoolcraft. Little Hill's narrative reflects the beliefs of the Winnebago Buffalo Society about the creation of the world.

Little Hill on Creation

The Great Spirit awoke as from a dream, and found that he was alone. He created the four winds by taking a piece of flesh from near his heart and mixing it with the substance upon which he sat. For these brothers he created a woman, our Grandmother the earth. She was sent down below, but she was unstable, and rocked about violently. To steady the world below, the Great Spirit sent down four giant snakes and four giant animals of another kind, and they were able to hold down the corners of the earth. However, when the winds blew across this creation, it fell back into unsteady motion again; so he created a gigantic buffalo, who is the land, and placed it in the center of the earth to make it steady.

ranges of ragged peaks ran along the length of the continent. The present-day plains (which for so long had lain along the west coast, exposed to the run of the sea) were now guarded by the serried ranks of the Rockies.

Early in this process, before the wall of mountains was complete, the ocean still sometimes slipped through gaps in the palisade and washed across the plains. This happened several times during the Triassic and Jurassic periods (between 250 million and 145 million years ago), culminating in a huge incursion, known as the Sundance Sea, which swept east into present-day Saskatchewan, Nebraska, and Texas. But these waters were soon expelled from much of their floodplain by a deluge of a completely different sort—an influx of mud and sand that washed down off the slopes of the newly formed mountains. No sooner had the mountains raised their heads than erosion began to level them. Mixed with generous quantities of volcanic ash from the tumult of mountain building, these sediments were strewn across the plains as far east as the Dakotas. Today they form brightly banded sandstones and shales—the Success, Kootenay, and Morrison formations by name—that bear witness to a titanic struggle among rivers, mountains, and seas. They also contain evidence of an awe-inspiring bestiary of ancient life.

The cataclysmic extinctions at the end of the Permian had left a biological void, but by the Middle Triassic (about 225 million years ago), this vacuum had been filled to bursting with reptiles. Creeping, crawling, swimming, flying, stomping across the land, reptiles had become the dominant animal group on Earth. Chief among them were the dinosaurs, including the 80- to 100-ton *Brachiosaurus*, which raised its ultralong neck to browse in the treetops, and the plated *Stegosaurus,* which had seventeen trapezoidal shields of bone embedded along its spine. Unfortunately for them (but fortunately for succeeding generations of dino-enthusiasts), hundreds of these large-bodied, small-brained animals apparently tramped into the rushing rivers, got stuck in the mud, and died. Their bones were then swept away by the currents and dropped on snags and in backwaters, where they lay in thick beds. These Morrison deposits provide the focus for the Dinosaur National Monument in Colorado and Utah. Although now in the mountains, the deposits formed on level terrain, and the same or similar species must have lounged under ginkgo trees and trudged through the spiky underbrush of what is now the Great Plains.

The sea, which in the Late Jurassic had been driven off the continent by sediment from the mountains to the west, managed to creep in one last time during the Cretaceous Period (between 145 million and 65 million years ago). By then, the Rocky Mountains formed an unbroken dyke along the west coast, but the rising waters rushed around it at both ends, flowing south from the western Arctic and north from the Gulf of Mexico. By the time the waters met in Colorado about 100 million years ago, almost the entire prairie region was inundated. It would take about 40 million years for the sea to make its final retreat, but when the water was finally gone, a vast plain lay exposed, stretching farther than the eye could see across the interior of the continent.

The first eyes to gaze across those broad, unfettered vistas were no doubt reptilian. By the Late Cretaceous, herds of heavyset *Triceratops,* with their wide, frilled collars and clustered horns, were roaming across the countryside and foraging in lush stands of horsetails, ferns, gingkoes, and palms, keeping an eye peeled for their most dangerous enemy, *Tyrannosaurus rex*. In case we were in any doubt about what *T. rex* preyed upon, a paleontologist in Saskatchewan once found a large sample of fossilized tyrannosaur dung. The two-quart (two-liter) lump contains what appears to be the crushed head frill of a juvenile *Triceratops*. Much of what we know about Cretaceous dinosaurs, both trivial and profound, comes from sites on the Great Plains, including the Red Deer River Valley in Alberta and the Frenchman Valley in Saskatchewan.

Tyrannosaurus rex

Succored by a mild and equable climate, much improved since Permian times, life in the Cretaceous was full. The air thrummed with insects, including moths and bees. The massive flying reptile *Quetzalcoatlus* rode the updrafts over the southern plains on a 40-foot (12-meter) span of wings—wider than those of a single-engine plane—searching for the bodies of the dead and dying. Frogs and salamanders hid in the underbrush, a habitat they shared with cowering, timorous mammals, few of them bigger than mice. Yet by the end of the Cretaceous Period, about half of this rich assembly of species—including all of the flying reptiles and the dinosaurs—had vanished. In fact, you could say they literally disappeared in a flash. One day, about 66 million years ago, an enormous meteorite (6 miles, or 10 kilometers, in diameter) crashed into the ocean off the present-day Yucatan coast with the force of a 100-million-megaton bomb, creating a worldwide holocaust of toxic vapor and soot. Beads of sizzling-hot glass were ejected from the blast, setting the rain forests and swamps alight as far north as Saskatchewan and Alberta. A newly discovered fossil site in southwestern North Dakota captured the catastrophe as it occurred: Charred trees. Shocked minerals. Marine and freshwater creatures all thrown in together. Fish with fused glass in their gills. By the time things eventually settled, the dinosaurs were extinct, and the plains were open for a new group of species to fill the vacancy.

HIGH AND DRY

NOT LONG AFTER the last dinosaur drew its last breath, something strange began to happen along the western margin of the Great Plains, in the heart of present-day Montana and Wyoming. About 50 million years ago, for reasons that no one can explain (more crashing and grinding off the west coast?), the level plains of the Cretaceous seabed began to heave upward, bend, and in places, crack open. Molten rock from the asthenosphere bubbled up through the fissures, sometimes crystallizing before it reached the surface, sometimes pouring out across the land to form dykes, domes, and ridges of lava. When the smoke cleared, mountains stood right out in the middle of the level plains. Subsequently honed by erosion, these unexpected rocky peaks still punctuate the western landscape from the Sweet Grass Hills east through the Bears Paw and the Little Rocky mountains, and south to the Crazy Mountains and the Black Hills.

At the time of their formation, the isolated "prairie mountains" did not have the presence that they do today. Even the main ranges of the Rockies were little more than bumps that protruded above a muddy, gravel-strewn landscape. The higher the mountains had thrust themselves up, the faster erosion had worn them away, until they lay buried, neck-deep, in their own shed silt, sand, rock, and clay. (The thick coal deposits in the Powder River Basin of northeastern Wyoming were formed when tons of this muck overran a peat bog some 50 million years ago and buried the vegetation under 10,000 feet, or 3,000 meters, of sediment.) Year after year, rivers carried a massive tonnage of this debris eastward to the central plains, depositing it as a broad, sloping alluvial fan. As the braided streams of the floodplain washed over the sediments, they gradually licked the surface smooth, creating a landscape that in places is so level that it almost seems supernatural. This stunning flatland once extended from the knob-peaked Rockies across southern Alberta and Saskatchewan, south through the eastern Dakotas, east to the Flint Hills of Kansas, and down to central Texas. Today, though much diminished by erosion, this landscape persists as the High Plains of Nebraska, Kansas, Oklahoma, and Texas. Legend has it that when the Spanish first crossed the plains of Texas in the 1500s, they used stakes to mark their route because the land was so spectacularly featureless. Hence the name Llano Estacado, or the Staked Plains, of northern Texas.

Some 45 million years ago, when the High Plains landscape was still being shaped, it would have taken more than stakes to help travelers find their way, for it was covered by a dripping, tangled forest. Globally, the climate had never been more amenable to life—there were dawn redwoods near the North Pole—and the plains basked in warm, wet, subtropical weather. A lush woodland spread across the midcontinent, alive with an impressive variety of birds and mammals. Ancestral squirrels and monkeys leapt through the overstory, while down below, titanotheres—beasts the size of rhinos, with knobby horns and sharp tusks—shuffled across the forest floor feeding on shrubs. Among the other browsing animals of the time was an early ancestor of the horse, *Orohippus* by name, which had four toes on its front feet and three on the back and grew to be about the size of a large Shetland sheepdog.

Life was easy. But then a sequence of unrelated events halfway around the world sent the climate into a nosedive. Beginning about 37 million years ago, the average global temperature dropped by 14°F (8°C) over the span of a million years. Thereafter, despite brief periods of recovery, the climate continued to cool. As the weather became cooler and drier, the tropical forests of the North American plains began to wither and die away.

But conditions that were death for palm trees were ideal for another group of plants. Relative newcomers on the evolutionary scene, grasses had first appeared shortly after the extinction of the dinosaurs but had met with limited success. They were drought specialists, and while humid conditions prevailed, they had been confined to small patches of ground that had somehow been deprived of abundant rainfall. Now, not only were the tropical rains failing because of a global drying trend, but the North American plains were under a special disadvantage. With the Rockies in place, storms that rolled in from the Pacific tended to drop their precipitation as they swept up the western slopes. By the time they reached the plains, they were pretty much wrung out. But grasses don't require much moisture, and this characteristic gave them a competitive edge. Over the next several million years (between about 24 million and 3 million years ago), grasses gradually became the dominant plants across the Great Plains.

If we could slip through a crack in time and go back to the plains of Nebraska some 20 million years ago, we would find ourselves in a landscape that is at once familiar and wonderfully strange. This is big sky country, an open landscape of shoulder-high grasses dotted with walnuts and other broadleafed trees, vaguely reminiscent of the savannas of East Africa today. A broad river courses across the plain, its margins fringed by willows and its current murky with sediment from the constantly eroding Rockies. Whenever this river floods, it coats the land with yet another layer of silt and sand.

The river is the main source of water in this increasingly arid land, and wildlife flocks to its banks. Herds of miniature rhinos (about the size of domestic pigs but with two horns sprouting from the ends of their snouts) plunge into the shallows to find refuge from biting flies. Ancestral horses called *Parahippus*, somewhat bigger than *Orohippus* but still the size of dogs, come down to the river to drink at dawn and dusk. The rest of the time, they range across the savanna, plucking leaves off the trees and grazing on grasses that tower over their heads. Because grass is very abrasive, *Parahippus* have acquired specially ridged teeth that are able to withstand the daily grinding. Llamalike camels (members of a family that evolved in North America and only later migrated to South America and Eurasia) lounge in the willows but keep an eye out for any suspicious shadows moving through the bushes. In this world, danger takes the forms of saber-toothed cats and long-jawed dogs, some of them as large as coyotes and wolves. Smaller dogs, the size of foxes, prey on the *Paleolagus*, or "ancient rabbits," that burrow into the roots of shade trees, and on *Paleocastor*, or "ancient beavers," that, amazing as it seems, occupy deep, corkscrew burrows in the middle of the dry prairie.

Dwarf rhinoceros

Traces of these animals, and others like them, have been preserved at the Agate Fossil Beds National Monument on the Niobrara River in northwestern Nebraska. Here, the buried beds of bone testify not only to remarkable lives but also to miserable deaths. It seems that the drying trend, which had driven back the rain forest and allowed the lush parklands to spread, occasionally became so severe that it stressed even the savannas, causing rivers to dry up and trees to blacken. Animals gathered alongside the dying rivers and died with them. Later, when floods flashed down out of the mountains, the currents gathered up the bones, massing them into backwaters and oxbows.

As the centuries ticked by, the climate became progressively more arid. Soon, in place of the lush savannas, a tawny, almost-treeless grassland sprawled across the plains. And although many mammalian species survived—including rhinos, horses, camels, rodents, cats, and dogs—all were challenged by their changed and unforgiving environment. An unremitting diet of grass pushed grazing animals to develop high-crowned teeth, which grew in to replace themselves as they were worn away. The absence of hiding places put a premium on speed, forcing both predator and prey to adopt the runner's long-legged physique. Hunter and hunted also came to rely on their quick wits, as the brain power of both players was augmented.

As it turned out, these hard-won adaptations would offer little protection against the trauma that was about to unfold—the Ice Age.

PERMANENT WINTER

NOBODY KNOWS FOR sure why the cold settled in as it did. Whatever the cause, by between about three million and two million years ago, the Earth had cooled so much that permanent winter had settled over the northern reaches of the continent. The tepid summers no longer melted away the preceding winters' snows. Beginning at high latitudes and progressing southward, drifts massed into mounds, and mounds into mountains, until the snow compacted into ice under its own tremendous weight. Eventually, after several thousand years, these glaciers began to advance, flowing almost imperceptibly but relentlessly south over the Central Lowlands. In time, the northern third of North America was buried under some 2 miles (3 kilometers) of ice; that's about the height, from base to peak, of Mount Everest. In its heartland on the Precambrian Shield, the ice reached a maximum depth of about 16,000 feet, or 5,000 meters.

Geologists used to believe that the glaciers advanced and retreated four times over a span of about two million years. These successive incursions were known in North America as the Nebraskan, Kansan, Illinoian, and Wisconsin glaciations, in honor of their southernmost extent. But more recent research suggests that the glaciers probably made many more than four sweeps down the continent, each time grinding away the traces left by previous glaciations. Since much of the record has been wiped clear, a detailed chronology of the Ice Age on the prairies cannot be reconstructed. But we do know that by about 1.2 million years ago, a vast slab of ice had bulldozed its way almost to the present-day confluence of the Missouri and Mississippi rivers. At its maximum, the ice sheet probably extended over the Canadian Prairie provinces and south through northeastern Montana, the Dakotas, and northeastern Kansas. From there it cut across the plains of northern Missouri and then eastward, across the continent, to the ice-stricken valley of the St. Lawrence.

After that ice sheet retreated, the glaciers never again penetrated quite so deeply into the plains. The final glaciation, for example, which began some 100,000 to 75,000 years ago, didn't progress much farther south than central Iowa. But the devastation that the glaciers inflicted was not limited to their actual footprint. Whenever the glaciers melted back, they left behind outwash plains of sand and silt. Ferocious winds that developed over the ice fields picked up this grit and hurled it around the interior of the continent. In a number of places (notably, the Great Sandhills of Saskatchewan and the Sandhills of western Nebraska) the wind laid down its burden in vast fields of dunes. Elsewhere, the storms whipped up clouds of dust—rock that had been ground into flour by the glaciers—and broadcast it over the land. Today, these silt, or loess, deposits, often several yards thick, form the bluffs along the Iowa side of the Missouri River and provide the matrix for rich, rolling farmlands in Nebraska, Kansas, Iowa, Missouri, and elsewhere.

The ice began its final, halting retreat about eighteen thousand years ago. Over the succeeding millennia, a block of ice larger than present-day Antarctica gradually melted away, and it didn't go quietly. Torrents gushed from the eroding ice sheets, gouging out meandering coulees and wide, flat-bottomed river valleys as they coursed eastward over the plains. Today, dry coulees writhe incongruously across the northern prairies, from nowhere to nowhere, and glacial spillways seem ludicrously oversized for the quiet streams, like the Milk River, that now occupy their broad channels.

And it wasn't only moving water that left its mark on the land. In many places, meltwater was prevented from flowing away by ice dams, and the silt-laden water pooled to form shallow, milky lakes, such as Glacial Lake Regina in south-central Saskatchewan and Lake Dakota in east-central South Dakota. The largest of these "proglacial" lakes, Glacial Lake Agassiz, flooded some 135,000 square miles (350,000 square kilometers) at its maximum extent (three times the size of Lake Superior, the largest modern freshwater lake), including extensive tracts in Saskatchewan and Manitoba and the Red River Lowlands in eastern North Dakota and northern Minnesota. When the ice and then the water finally retreated from the land, these lake bottoms stood exposed as broad plains bounded by terraced beaches, all covered with a dressing of mineral-rich silt.

The land that emerged directly from under the ice sheets, by contrast, was a rough-and-tumble mess, strewn with the rubble that the glaciers had dropped as they retreated. Sinuous ridges of gravel and silt, called eskers, marked the courses of streams that had once flowed under or through the ice; strange conical hills, called kames, stood where streams pouring out of the glaciers had deposited gravel and sand. One of the most prominent glacial features on the northern plains was a long, broken ridge of hill country, called the Missouri Coteau, that meandered (and still meanders) across central Saskatchewan and south through the Dakotas. Geologists refer to the Coteau as "dead ice moraine," because it formed when hunks of ice became buried in gravel and lay there for centuries, ever so gradually rotting away. As each block of ice melted, the gravel that had been lying on top of it sagged to form a depression, or prairie pothole.

Meanwhile, south of the reach of the glaciers, on the foreshore of the Rockies, the landscape had also been undergoing renovations. Sometime before the Ice Age set in, the entire western prairies had inexplicably begun to rise. As a result, the rivers, which previously had been building up the plains with loads of gravel and silt, now began to cut through the very layers they had previously deposited. This erosion was most dramatic along the slopes of the Rockies, where the rivers were powerful enough to wear through 70 million years of sediments. Along the Front Range of the mountains in Colorado, for example, the South Platte and Arkansas rivers have dug 1,600 to 2,000 feet (500 to 600 meters) below the level of the High Plains. Only where erosion-resistant layers of rock have stood against this assault can the remnants of the older landscape be seen. The tops of the buttes and mesas that tower over the eroded landscape were once a part of this continuous high plain.

By the end of the glaciation, the Great Plains of North America had been transformed from the seabed of ancient times into a mosaic of distinctive landforms. To the north extended a rumpled terrain of glacial debris. Beyond the limit of the glaciers, to the south and east, lay a softer landscape of ancient ocean floors, much of it now blanketed in wind-shaped drifts of glacial sand and silt. To the west, the flatlands of the High Plains stepped up steadily toward the front ranks of the Rockies. And everywhere, rivers were cutting down into the land, etching deep valleys, canyons and, where the land was suitably dry and bare, badlands.

But if the varied landforms of the plains were beginning to look more like those of the present, many of the life-forms still did not. Disadvantaged by the cool, wet weather of the Ice Age, the grasses that had previously dominated the plains had lost ground to other plants. Now a band of tundra skirted the retreating ice, while to the south, dark coniferous forests spanned much of the continent. Pure grasslands were restricted to scattered meadows and, perhaps, to a relict prairie crammed into the southernmost plains. Together, these diverse habitats were occupied by a stunning array of life, including white-tailed and mule deer, caribou, several species of pronghorns, black bears, cougars, bobcats, lions, cheetahs, saber-toothed cats, horses, llamas, one-humped camels— even Ice Age elephants. Woolly mammoths (shaggy beasts that stood almost 10 feet, or 3 meters, tall) browsed on the tundra, while Columbian mammoths (just as unkempt and much larger) appear to have favored the remnant patches of grassland. Meanwhile, in the forests, their somewhat daintier relatives, the mastodons (the size of Indian elephants) fed on a diet of black-spruce boughs and other woody tidbits.

Columbian mammoth

The mammoths and mastodons were relatively recent arrivals on the plains, Ice Age immigrants that migrated across the Bering land bridge from Eurasia during intermittent cold spells. Whenever the climate worsened and the glaciers advanced, water became locked up in the ice and sea levels dropped, exposing a bridge of land across the Bering Strait between Siberia and Alaska. When the glaciers receded again, the land bridge was drowned, but a passageway simultaneously opened to the south through the Canadian plains, which allowed the newcomers to wander into the heart of the continent. Some mammals, including ancestral camels and horses, made this journey in reverse, moving north when the plains corridor was open and then migrating across to Asia when the land bridge appeared.

Of all the species that arrived on the North American plains during the Ice Age—a menagerie that included not only elephants but also grizzlies, elk, and moose—two demand special notice. The steppe bison, *Bison priscus*, was a magnificent, thick-maned animal with flamboyant curving horns (attributes that are dramatically depicted in the cave art at Lascaux, France). The first bison herds likely poured across the Bering land bridge a few hundred thousand years ago and eventually made their way south to the Great Plains. Over the millennia that followed, successive waves of steppe bison made the same long trek, eventually meeting and mingling with the descendants of the pioneer herds. Meanwhile, that pioneer stock had been changing, shaped by life on the steppes and forests of a new continent. The result of this complex process of immigration, adaptation, and interbreeding was the emergence of several distinctively North American types, notably the giant, long-horned *Bison latifrons* and the somewhat smaller *Bison antiquus*. In time these species were displaced by an even more compact version, *Bison bison*, the shaggy beast that, in historic times, provided food and shelter to the first people of the plains.

Exactly when the first of those hunting people arrived on the scene is a mystery. Until quite recently, most archeologists insisted that humans (members of a genus that was born in East Africa some two million years ago) entered North America from Asia, by crossing the Bering land bridge and traveling down an ice-free corridor into the plains. This migration was said to have happened about 13,000 years ago.

Then, in the 1970s, researchers working in Chile uncovered evidence that people had been living there for two thousand years before this supposed first-arrival date. Subsequent discoveries in Bell County, Texas, and elsewhere have pushed the timelines back even further, to at least 15,500 years before the present. This accords with the spirit of Indigenous memory, which affirms that their forebears were here at the beginning. Equipped with elegantly chipped fragments of stone and bone, these ancestral hunting people killed and butchered not only bison but also camels, horses, mastodons, and—their specialty—mammoths. At sites from Alberta to Texas, the proof of their presence—blackened hearths, discarded tools, and cracked marrow bones—lies buried where they left it so long ago. In some places, the skeletons of several large mammals lie strewn about the camps, testimony to the prowess of these big-game hunters.

But inevitably, on a planet where change is the only constant, this regime was fleeting. By thirteen thousand years ago, the fabulous array of large mammals on the plains was already disappearing. As many as fifty species—including giant beavers, ground sloths, lions, cheetahs, dire wolves, saber-toothed cats, horses, camels, mammoths, and mastodons—all became extinct within a few thousand years. Did an exploding population of well-fed humans hunt the animals into oblivion, as some archeologists believe? Or was climate change the culprit? The evidence suggests that, between about thirteen thousand and ten thousand years ago, average global temperatures first dropped abruptly and then rebounded. On the North American plains, these climatic changes ultimately translated into a dramatic shift in vegetation patterns. Pushed by warmer, drier conditions, the spruce forests gave way to pines, then in places to open, mixed woodlands, and ultimately to grass. In the blink of an eye (geologically speaking), a carpet of grasses spread out across the plains, blued by sage and beardtongue and enlivened by patches of golden beans, blazing stars, and prairie smoke, or three-flowered avens. The prairies of historic times had finally been created.

This new grassland was big and bold, but it was also much less varied than the mixed landscape of tundra, grass, and forest that it had displaced. And perhaps this in itself is enough to explain the disappearance of the Ice Age megamammals, which required a rich and varied supply of foods that grasslands alone could not provide. Yet if the new habitats were insufficient to sustain mammoths, they appeared perfectly suited to the bison, which soon emerged as the dominant grazing animal on the open range.

As the great herds drummed across the prairie, they of course had no conception of how their homeland had been shaped. The evidence of that drama lay unnoticed beneath their hooves—long-buried evidence of onrushing seas, rising mountains, silt-burdened rivers, and towering cliffs of ice. But the bison were untroubled by the traumas of the past, as they flowed across the horizon toward the present.

CHAPTER 3

THE GEOGRAPHY OF GRASS

I am the grass.
Let me work.

CARL SANDBURG, "GRASS," 1918

G RASSES ARE THE heart and soul of the prairie, the living link between the physical resources of the Great Plains—sunshine, rain, and soil—and almost every other aspect of the ecosystem. At first glance, grasses may look simple or even primitive. In fact, they are highly evolved organisms, especially adapted to cope with extreme climatic uncertainties, including frequent drought. From probable origins in the African region of the supercontinent Pangaea some 60 million years ago, grasses have migrated to every continent except Antarctica and have diversified into about 10,000 species throughout the world. Of these, approximately 140 species, in 41 genera, naturally occur in the Great Plains Grasslands. That's nearly twelve dozen distinctly different native grasses! Some of them, like the magnificent big bluestem, or turkey foot (so called for its large, three-lobed seed head), grow up to 10 feet (3 meters) tall. Others, like the stick-in-your-socks specialist needle-and-thread grass, seldom exceed 3 feet (a meter) in height. At the low end of the scale are species like the diminutive blue grama, which grows close to the ground and rarely raises its elegant, eyebrow-shaped seed heads more than a few hand widths above the soil.

Yet despite these obvious differences, the prairie grasses all share one crucial ability. They are tuned in to the climate, able to dial their metabolisms down when conditions are unfavorable for growth and speed them up when the weather improves. Far from being passive stalks blowing idly in the wind, prairie grasses are lean, mean growing machines, designed to make the most of limited and unreliable resources.

MANAGING MOISTURE

ONE KEY TO the prairie grasses' success is their ability to conserve water. Like most plants, grasses take in water through their roots and lose it as water vapor through tiny mouth-shaped valves, or stomata, in their leaves. The larger the surface of the leaf and the more stomata it bears, the greater the risk that the plant will lose too much moisture through evaporation, causing it to collapse. Grasses are protected from this trauma by having a reduced number of stomata and by the design of their leaves, which take the form of narrow blades. What's more, the surfaces of these reduced leaves are often modified—corrugated with ridges or covered in hairs—so that the wind can't sweep across the surface and draw out moisture. The roughened surface holds a thin layer of humid air next to the leaf and thus helps to reduce the "evaporative demand," or drying power, of the atmosphere. Some grasses, including western wheatgrass, June grass, and blue grama, roll up the edges of their leaves during times of drought to help keep their tissues from drying out.

Why aren't the stomata kept tightly closed to seal moisture inside the leaf? The reason is that the stomata also supply plants with fresh air. Leaves are miracle workers, able to take carbon dioxide from the air and water from the soil, zap them with solar energy, and transform them into food. This process—photosynthesis—not only produces the sugars and other organic molecules that plants need to maintain themselves and to grow but also feeds microbes, worms, insects, fish, birds, and mammals. If plants sealed their stomata, this life-sustaining process would come gasping to a halt for lack of carbon dioxide. But if the stomata are thrown wide open, the plants risk death due to the loss of moisture through their gaping valves.

Prairie grasses resolve this dilemma by strategic scheduling. In the fierce blaze of the midday sun, the stomata close so that water vapor is held in and carbon dioxide is kept out. In this state, the leaf can capture solar energy and store it in energy-rich molecules (a process that requires sunlight but not carbon dioxide). Then, in the cool of the evening, when the evaporative demand drops off, the stomata snap open, letting water vapor trickle out but also permitting carbon dioxide to flood into the leaf. By mobilizing the energy that was stockpiled earlier in the day, the leaf uses this carbon dioxide to manufacture the sugars and other molecules that it needs for growth (a process that can be accomplished in total darkness). The result is that prairie grasses are partially nocturnal; they do most of their growing at night or in the early hours of the morning.

Prairie grasses also have another ingenious way of evading the demands of the sun. Like many other grassland creatures (prairie dogs, ground squirrels, cottontails, badgers, and so on), they take refuge underground. What we think of as "grass"—the aboveground leaves and stems—actually constitutes less than half of the organism. Between 60 and 80 percent of the plant, by weight, typically grows below ground. The roots extend down from the base of the stems like a tangled head of hair, as main roots divide into minor roots and then into root hairs. A 10-foot (3-meter) stand of big bluestem is anchored underground by a mass of coarse, fibrous roots that reaches as much as 12 feet (3.6 meters) into the earth. Blue grama, for its part, seldom lifts its seed heads very far above the ground, but its network of fine, branching roots can sometimes probe the soil for water to a depth of almost 6 feet (1.8 meters)!

These extensive systems of roots push thirstily through the soil, intent on sucking up every available drop of water. But if the soil is very dry, as it is during periods of drought, the roots can't draw in enough moisture to keep pace with losses from the stomata. Grasses respond by transferring their most valuable resources (including sugars and proteins) from their leaves into their roots and, especially, into their rhizomes—those aggressive, underground stems that are familiar to anyone who has ever battled with quack grass in the garden. Dead to the world above ground—withered and crisp—the plants live frugally below the surface, drawing on their cached supplies and biding their time until the weather improves. When the rains eventually return, as inevitably they do, the grasses explode into action, sending out fresh rhizomes, which in turn put out fresh leaves and roots, to produce a burgeoning network of tender growth. The amazingly resilient blue grama can revive from dormancy, green up, and grow on as little as 0.2 inches, or 5 millimeters, of rainfall.

Indian grass *Needle-and-thread grass*

Galleta *Western wheatgrass*

YOU CALL *THAT* A DROUGHT?

The Great Plains Grasslands, and the climate that defines them, have been around for the last eight thousand to ten thousand years. In the early days of this regime, the climate was considerably warmer and drier than it is today and even more prone to drought. But sometime in the last few thousand years, the system took a turn toward cooler, moister norms, so droughts have gradually become less frequent.

In fact, it seems that the twentieth century was the wettest in two thousand years. This conclusion is based on studies of microscopic fossils found in lake beds across the northern plains, in Alberta, Saskatchewan, and North Dakota. By extracting core samples from lake bottoms and studying the fossils that are found at different depths, researchers are able to estimate the salt content of the water at different times in the past. Since salinity increases as water levels drop, these findings give them a measure of past droughts. At Humboldt Lake in central Saskatchewan, for example, the fossils bear witness to a severe drought that persisted unbroken for more than seventy years. On the southern plains and in the desert United States, researchers have uncovered evidence of prehistoric droughts that lasted for three centuries.

Could the prairie climate revert to its fierce old habits? Yes, and it may already be happening. In 2017, for example, Amarillo, Texas, went 126 days without measurable precipitation, far surpassing a long-standing record. To make matters worse, average annual temperatures are on the rise—already up by 3.4°F (1.9°C) in some localities over the last hundred-plus years—as the prairies ride the leading edge of global warming.

Prairie grasses are not all equally capable of coping with drought. In general, tall grasses, including big bluestem and other shoulder-high species such as switchgrass and Indian grass, require the most moisture, while short grasses like blue grama, galleta, and the stubby little buffalo grass are the most resistant to drought. Midheight species, including needle-and-thread grass, rough fescue, and western wheatgrass (a.k.a. bluejoint, for its bluish leaf nodes), tend to fall somewhere in between. But all prairie grasses can contend with drought more successfully than can most deciduous trees—which is why the prairies are prairies instead of forests. The grasslands are an expression of the drought-prone prairie climate and a living response to the geography of the midcontinent.

WEATHER MATTERS

TO THE HOMESTEADERS who came to the Great Plains from Europe or eastern North America in the late 1800s and early 1900s, converting the prairies to cropland must have looked like a dream. Except for the trees that crept in along the rivers, the land lay open to the plow, offering little apparent resistance to the farmers' ambitions. But the settlers' early optimism was soon blighted by widespread droughts, as the dry summer of 1889 was followed by the dry years of 1890, 1894, 1910, and 1917, and then by the bleak decade of the 1930s. Life on the prairies was not as easy as it had seemed. For what no one at first quite realized was that grasslands are semiarid zones—better watered than deserts but less humid than forests. The farmlands that the settlers had known in

Europe and the East had typically been wrested from the forest and, even after the trees were gone, still received enough rainfall to support a natural vegetative cover of broad-leaved woodlands. But the weather on the prairies naturally favored not trees but grass, and that simple fact made all the difference.

Like most of the world's great grasslands, the Great Plains of North America lie squarely in the middle of a large continental landmass. As a result, the region is isolated from the influence of all four oceans—north, south, east, and west—and, as it happens, from any other significant body of water. Without the moderating influence of water (slow to heat and slow to cool), the plains are subject to violent oscillations of temperature. In the northern prairies, in particular, the temperature can span 140°F in the course of a year, from a brittle –40°F in midwinter to a stifling +100°F in summer. (That's a range of 80°C, from a low of around –40°C to a high of over +40°C.) The effect of these wild seasonal deviations is equivalent to moving up and down the continent every twelve months. Saskatoon, for example, has an average January temperature of 0°F (–18°C), well below that of Anchorage, Alaska. But in July, Saskatoon's average heats up to 66°F (19°C), almost on a par with Los Angeles. And though the southern plains are spared the worst extremes of winter, they still get taken for quite a ride. The average January temperature in Amarillo, Texas, for example, is a mere 35°F, or 2°C, cooler than Vancouver far to the north; but in July, the north Texas plains can be among the hottest places on the continent. (If prairie people are obsessed by the weather, it may simply be because we have a lot of weather to obsess about!)

Prairie grasses ride this climatic roller coaster with composure. Species that couldn't stay the course fell into extinction in ages long past, leaving the modern community of hardy survivors. Each of these successful species has been further refined over the past several thousand years, producing subspecies, or varieties, that are finely attuned to local conditions. A native grass from Alberta, for example, typically achieves maturity in a matter of weeks, fitting its life cycle to the abbreviated growing season of the northern plains. But a clump of the same species from Missouri or Oklahoma is programmed to take its time, pacing its activities to the more leisurely schedule of southern climes. These kinds of local, genetic differences have been detected in a wide range of native grasses, including blue grama, its cousin sideoats grama, the compact and graceful June grass, switchgrass, and both big and little bluestem. Natural selection, that master gardener, has been at work on them.

The one climatic factor that presents a continuing challenge for prairie grasses is the moisture supply. Over the Great Plains as a whole, precipitation is more variable than it is almost anywhere else on the continent, with years that are both much wetter and much drier than the long-term norm. (One study of precipitation records in western Kansas, for example, showed that in most months the amount of moisture received was either significantly below or significantly above average. Only "normal" values were truly abnormal.) When the rains are generous, the prairie flourishes and blooms; but when drought sets in, the grasses—indeed the whole ecosystem—are severely tested. More than any other single factor, the limits to growth on the prairies are set by precipitation.

PLANTS FIGHT BACK

PRAIRIE PLANTS HAVE come up with many ingenious strategies for coping with water shortage. A few, like the pincushion cactus, are genuinely drought resistant. In other words, they can store water in their own tissues (in their enlarged stems) and draw on it as needed. Others, including many grasses and wildflowers, attempt to evade drought by going dormant and retreating underground, where they linger on in the form of seeds, rhizomes, or tubers. But if some plants favor patient waiting, others put their faith in speed. Instead of trying to sit out the drought, they attempt to avoid it entirely.

Take, for example, the prairie crocus, or pasque flower. An inexhaustible source of pleasure for people on the northern plains, crocuses appear on the trailing edge of winter as tight clusters of furry, pointed buds that push up through the dead grass like so many inquisitive snouts sniffing for spring air. Without pausing to grow leaves, the plants burst directly into bloom, producing ground-hugging whorls of silky, lavender sepals. By the time most other wildflowers put in an appearance several weeks later, crocuses are already sporting headdresses of shiny, plumed seeds. Before the growing season has even properly begun, their reproductive task has been completed.

By getting off the mark so early, crocuses are able to draw on a relatively certain supply of water from snowmelt. And although they are exposed to the bluster of winter's last blast, they are protected from the wind by a coat of hairs that holds in heat and moisture. They also take shelter by crouching close to the ground, well bedded in grass, creeping juniper, and other plants. Thus protected, crocuses speed through their reproductive cycle and avoid the stress of coping with drought in the hot, dry days of July and August.

WHY SO DRY?

THE GREAT PLAINS are subject to drought partly because they lie in the lee of the western mountains. Without this elevated barrier, westerly winds from the Pacific could sweep across the plains and bring moisture to the dry lands from Airdrie to Abilene. But with the Coast Ranges, the Cascades, the Sierras, and the Rockies all standing in their path, the Pacific westerlies are forced to rise, cool, and drop their moisture as they pass. By the time the winds flow down over the plains, they are almost devoid of rain. As they move across the western grasslands, they pick up humidity and carry it to the well-watered eastern forests. (Sometimes, as the winds swoop over the mountains, they whip themselves up into disturbances—known as Alberta Lows or Colorado Cyclones, depending on where they occur—that sweep across the plains, carrying a splattering of rain or a dusting of snow.)

Once the westerlies get past the mountains, there is nothing to stop them—except for the invisible resistance of other air masses. The prairies' open spaces are a playground for the winds, drawing in not only mild, dry air from the Pacific but also colder and even drier air from the tundra and polar seas. As this Arctic air floods south, it meets moist, tropical air flowing north from the western Gulf of Mexico. A typical weather diagram for the Great Plains would show Arctic air pushing down from the north, tropical air swinging in from the south, and wrung-out Pacific air wedged between them like the point of an eastward-facing arrow. Arctic air dominates in the winter, sometimes forcing itself all the way to Texas as a stinging "blue norther" and, occasionally, pushing on south to the isthmus of Tehuantepec at the very tip of Mexico. In the spring, the balance of power is reversed, as the gulf air mass gains strength and surges north, sometimes carrying tropical heat and humidity all the way up to the Canadian prairies. The rained-out Pacific westerlies, with their meager stock of moisture, make themselves felt throughout the year, especially during the dry months of fall and winter. *See Map 5: Major Air Masses Affecting the Great Plains.*

In spring and summer, in particular, this picture is complicated by an influx of warm, dry-as-bone air that blows across the southern plains from the southwestern deserts. The interplay of these "four strong winds" produces the distinctive precipitation patterns of the Great Plains Grasslands. For instance, winter is a relatively arid season across most of the plains not only because cold air cannot hold much moisture but also because of the strong seasonal influence of dry air from the north and west. As a rule, less than one-third of the year's precipitation falls between October and March, when these air masses exert their strongest influence. The other roughly 70 percent of the year's moisture is received during the April-to-September growing season. Without this well-timed gift, the Great Plains would be a prickly expanse of cactus and other desert plants.

Much of the all-important spring-and-summer rainfall is generated when tropical air surges north and runs into Pacific and Arctic systems moving across the prairies to the east and south. Where the air masses collide, the lighter, warmer air from the tropics is forced up, cooling as it climbs, condensing to form clouds, and ultimately losing its moisture as general rain showers. Violent thunderstorms also frequently develop along these collision zones, or fronts, as the unstable tropical air rises up into towering, super-energized cumulonimbus clouds that glower over the landscape before releasing their humidity as hail or pounding downpours.

Because the contending air masses often meet in mid-continent, frontal thunderstorms are most common in the middle of the plains, in and around Colorado, Wyoming, South Dakota, Nebraska, and Kansas. But storms can also develop locally, without the clash of opposing weather systems to set them off. All it takes is a mass of warm, moist air and something to send that air spiraling up through the atmosphere. This lift-energy usually comes from the summer sun, which blazes down through cloudless prairie skies to heat the ground. Heat then radiates out of the soil into the surface air, causing it to rise, rotate, and mount upward to form a rain-filled thunderhead. In the dry western plains, one-third of the year's precipitation can fall in a single hour from one of these spectacular cloudbursts.

GLOBAL "TELECONNECTIONS"

BY AND LARGE, the prairie climate is reliably unreliable. As the rival air masses interact with each other over the plains, they keep the atmosphere in a state of more-or-less-constant flux, so that the weather oscillates from extreme to extreme. But there are also times when the climatic system seems to get stuck. "Wet spells," for example, when the rain refuses to stop. "Dry spells" of months—or years—when the clouds seem dry as parchment and the air fills with dust.

These persistent weather patterns also tend to be widespread, affecting significant parts of the Great Plains for prolonged periods. The droughts of the 1930s, for example, occasionally flared out to singe the entire continent, but they were at their most intense across the Great Plains Grasslands. Some parts of the High Plains in Oklahoma and Texas experienced eight consecutive years of drought, between 1933 and 1940. Little more than a decade later, the central and southern plains—from the Mississippi to the Rockies and from Colorado to Texas—were again stricken by a severe drought that persisted from 1952 to 1957. The Canadian Prairie provinces were hit hard in 1961. Then, in the late 1980s, a three-year drought parched the entire northern plains and fueled disastrous forest fires in Yellowstone National Park. During the growing season of 1988, when the crisis was at its worst, many parts of the prairies were hotter and drier than they had been at any time during the Dirty Thirties.

Yet five years later, some of these same areas were in full flood, as torrential rains pounded the western Midwest and sent both the Missouri and the Upper Mississippi rivers spilling over their banks. By the time the waters receded, twenty-six people were dead.

Why do the prairies suffer these violent climatic spasms? Part of the answer to this question may lie halfway around the world, in a region somewhere between Australia and Peru. There, in the equatorial waters of the South Pacific Ocean, weather patterns that will eventually affect the prairies begin to brew. Recent research suggests that there is a link between the surface temperature of the South Pacific and the amount of precipitation that ultimately falls on the Great Plains, particularly during the winter and early spring.

So far, no one knows exactly how all the complex linkages in this world-wide "teleconnection" work. And global influences, however stupendous, are not the only factors involved. Often, extreme conditions linger on the Great Plains long after the systems that triggered them have dispersed. A wet spell seems to breed more wet weather; a dry spell appears to breed more drought. But how could weather patterns possibly perpetuate themselves? The answer turns out to be surprisingly obvious. When precipitation is plentiful, water accumulates in the soil. As plants draw on this moisture to grow, they release water vapor into the air. This water vapor, in turn, combines with humidity that has evaporated directly from the earth, and these exhalations rise together to form clouds. Thus rain in the soil begets rain showers. What's more, both rainfall and evapotranspiration (the release of water vapor from plants) have a cooling influence that helps to moderate temperatures and keep the evaporative demand within comfortable limits.

After a prolonged dry spell, by contrast, the cycle grinds to a stop. Plant growth slows and the rate of transpiration declines. So too does cooling evaporation from the soil. The ground and the surface layers of air sizzle in the sun, as a hot, dry land gets hotter and drier. (A case in point is the drought of the 1930s, which seems to have been intensified and prolonged by farming methods that left the soil exposed to the parching wind and robbed the system of what little moisture it held.) Eventually humid air from the south or the west returns to the scene, bringing welcome relief and restoring the climate to its own eccentric sense of normalcy.

GRADIENTS OF GRASS

THESE PROLONGED EPISODES of drought have been the making of the Great Plains Grasslands. Drought sucks moisture out of the soil, beginning at the surface and gradually burning farther down. If the dry spell is brief, the deep stores of moisture remain untapped, but if the evaporative demand persists, even the subsoil becomes parched and cracked. As a result, deeply rooted trees can cope without rain for several years by drawing water from underground, but they are doomed to defeat when drought reaches their root zone. Meanwhile, the grass lies patiently around the dying trunks, ready and able to spring back to life when the new rains finally come.

Long-term patterns of precipitation not only determine whether the land will grow trees or grass but also establish the limits that distinguish one "type," or ecoregion, of grasslands from the next. When precipitation and other variables are averaged over the long term, the underlying order of the prairie climate begins to emerge. In the textbooks, these hidden patterns are revealed through charts and maps, but out on the prairie, they are written as gradients of grass. *See Map 6: Average Annual Precipitation on the Great Plains.*

Sometimes, the dialogue between the vegetation and the climate is intriguingly complex. For instance, summer precipitation on the prairies depends, in large part, on air masses that blow in from the south, carrying moisture from the gulf.

PRAIRIE FIRE

Climate is the major factor that determines the extent of the Great Plains Grasslands. Technically speaking, grasses hold sway when the evaporative demand (the amount of moisture that the atmosphere would draw away if it could) is slightly greater than the precipitation (the amount of moisture that is out there, in the ecosystem). But there is one important exception to this rule. The lush tall-grass prairies that fringe the eastern margin of the plains receive abundant moisture, more than enough to keep pace with evaporation. Theoretically, the region ought to support trees. And, in fact, wherever fragments of tall-grass prairie have survived, they have been aggressively invaded by stands of aspen, oak, and dogwood during the last 150 years.

The missing link is fire. Prairie fire was the terror of the early settlers, but it was a friend and ally of the tall grasses. Not only did it clear away the thatch of dead vegetation that prevented new shoots from breaking through, it also killed trees, the true "terror" of the prairies. When a tree burns, the growth points on its twigs and branches are often injured, so the plant cannot easily produce new shoots. But a grass protects its growing tips under the ground and rises from the flames like the proverbial phoenix.

Before the agricultural era, most of the tall-grass region probably burned every three to ten years, set ablaze by lightning or by Indigenous people, who used fire to green up the prairie and bring in animals. But however the flames were ignited, they had the same effect: they renewed and sustained the tall-grass prairies.

Because of their southerly origins, these winds naturally have a greater influence on the southern plains (where they "reside") than in the north (where they merely "visit"). So it isn't entirely surprising to discover that the southern plains receive significantly more moisture than the northern prairies do. If, for example, Amarillo can hope to get 20 inches, or 500 millimeters, of moisture in a normal year, Lethbridge typically has to make do with only three-quarters as much. With this difference in mind, one might expect the prairies of northern Texas to be lusher than those of southern Alberta or Saskatchewan. Instead, the reverse is the case.

The trick is that the south-to-north gradient in precipitation is canceled out by an equal but opposite north-to-south gradient in evaporation. Because the average annual temperature increases from north to south, so does the rate at which moisture is lost through evaporation. Whatever the southern plains gain as rain, they lose as water vapor. As a result, the "effective precipitation"—the amount of water that is available to growing plants—is about the same in southern Alberta as in northern Texas. This helps to explain the long, gradual transition from the semiarid climate of the Northwestern Short/Mixed Grasslands to the sun-frazzled conditions of the Southern Short Grasslands. (As climate change imposes hotter, drier conditions on the southern plains, this gradient is expected to become even more dramatic.) *See Map 4: Ecoregions of the Great Plains.*

Meanwhile, there is yet another climatic gradient that helps to shape the vegetational profile of the Great Plains. This is an east-to-west decline in average annual precipitation. The tropical air that brings summer rains to the prairies typically swings up from the Gulf of Mexico, through the central

United States, and off toward the east. As a result, its influence is stronger on the tall grasslands of the eastern plains than on the short-to-mixed grasslands farther west. If Winnipeg receives about 20 inches (500 millimeters) of moisture on average, Lethbridge gets 20 percent less (just 16 inches, or 400 millimeters). And if Kansas City can count on 40 inches (1,000 millimeters) of precipitation in a normal year, Amarillo can only expect to receive about half as much—and this time there is no reverse gradient in temperature to compensate for the difference. Less moisture is simply less.

In the days before the prairies were plowed and settled, this east-west moisture gradient found expression in the natural vegetation. As the average precipitation declined toward the west, the vegetation diminished in step, gradually reducing in height like a living bar graph. Somewhere around 100 degrees west longitude (give or take a few degrees), the tall-grass species dwindled away, leaving the wind to ripple through knee-high stands of mixed-grass prairie. To the west and southwest, the midheight grasses in turn gave way to a carpet of ground-hugging grasses, as a dry land clothed itself in drought-resistant plants.

The boundaries between tall-, mixed-, and short-grass prairies are not as tidy as they look on the map. In the patchwork quilt of the grasslands, each of the major blocks of vegetation is composed of many smaller blocks. And just as the overall picture is determined by large-scale climatic patterns, so each of these distinctive patches is a response to local variations in microclimate. Conditions are different on a south-facing slope than on the north and on lowlands than on hilltops, and these subtle differences are reflected in the vegetation. Moving uphill, from humid bottomlands to the drier crest, reproduces the moisture gradient of moving from east to west. And so, a sea of tall-grass prairie is broken by islands of mixed (or even short) grasses that grow on uplands and arid slopes. Meanwhile, out on the short-grass prairie, blue grama and its diminutive associates follow the opposite trend, ceding ground to midheight or tall grasses in moist valley bottoms.

Whether considered locally or regionally, the transition zones between the different "types" of prairie are dynamic, migrating upslope and downslope, east and west, following changes in the climate. During periods when the weather is unusually wet, the tall and mixed grasses expand their influence and range, tracking the sweep of abundant rain. In times when the prairies are stricken by drought, the tall grasses are forced into retreat and the drought-resistant short grasses extend their reach. (During the Dirty Thirties, for example, the short-grass community pushed as much as 150 miles, or 240 kilometers, to the east, temporarily usurping the tall-grass prairie.) Thus, the grasslands keep time with the climate, mirroring its ebbs and flows and dancing to the erratic beat of the rainfall.

No one can say for certain how this ancient partnership will adapt to the climatic uncertainties of the near future. With the planet now blanketed in a thick layer of heat-trapping pollution from human sources, average annual temperatures are rising around the world, and the Great Plains are no exception. In the United States, the fastest temperature rise documented to date was observed in North Dakota. Across the northern plains in general, there is a trend toward heavy precipitation events during the growing season, potentially causing floods, and intense heat waves in the summer, increasing the chance of drought. The outlook for the southern plains is even more unnerving, with a call for extended dry spells with temperatures above 100°F (40°C) and periodic "rain bombs" that cause sudden flooding.

Fortunately, one of the most promising ways to reduce the intensity of climate change is to protect and restore grasslands. Because prairie grasses put so much of their life force into their massive networks of roots, they not only withdraw carbon dioxide from the air but actually bury it, in the form of organic matter, deep below the surface. What's more, these spongy, humus-rich soils also absorb and hold moisture, helping to mediate between too much precipitation and too little. Although forests are rightly prized for their role in sequestering carbon, prairie grasses may yet prove to be the real superheroes of climate protection.

CHAPTER 4

SECRETS
OF THE SOIL

... the prairies are nothing but grass as
the sea is nothing but water.
WILLIAM LEAST HEAT-MOON, *PRAIRYERTH,* 1991

T HE GRASSLANDS OF the Great Plains tug at our senses, challenging us to overcome our limitations as mere humans. On the one hand, the landscape calls us to enlarge our horizons and soar across its spacious expanse. But on the other, the prairie seduces us with details, quietly drawing our attention to the little things that matter in this big, round world. The flick of a ground squirrel diving into its burrow, the wheeling shadow of a hawk, the ripple and run of its wingspan across the bending grass. If we are willing, the prairies will even lead us beyond these small sightings to truly minuscule events, beckoning us into an alien but all-important world that lies almost entirely beyond the reach of our ordinary awareness. For the typical life-form on the grasslands is not a bison or a gopher or even a blade of grass. It is some weird little creepy-crawly, far too small to be seen by the unaided eye, that lives and dies just beneath our feet in the deep prairie soil.

Recent estimates suggest that the total weight, or biomass, of the invisible organisms that live in prairie soils is greater than the mass of all the visible, aboveground animals put together. If all the microflora and microfauna from grassland soils could be piled onto one arm of some Great Cosmic Scale, with all the insects, reptiles, mammals, and birds from the upper world heaped onto the other pan, the balance would tip sharply toward the soil organisms. Together with the plant roots around which they live, these little creatures are the powerhouse of the prairie, responsible for anywhere between 60 and 90 percent of all the biological activity in the Great Plains Grasslands.

ENTERING THE UNDERWORLD

THE SOIL HAS been described as the "poor person's tropical rain forest," one of the richest and most diverse ecosystems anywhere on Earth. The most numerous organisms in this busy world are also the smallest—thousands upon thousands of different kinds of single-celled organisms, including algae, fungi, bacteria, and protozoans. Although a few of these creatures may be familiar from Biology 101—the ooze-along, blob-shaped amoebae, for example, and the whip-tailed flagellates—many (if not most) of the soil microbes are still completely unknown, never having been studied, named, or even seen by anyone. A single teaspoon of dirt typically holds around five billion of these fantastically varied little animals and plants, all of them fiercely engaged in the business of life and death. In their billions and their trillions, they expend as much energy as if several thousand people were living on every acre, or hectare, of the prairies.

Adding to the hubbub is a lively assemblage of somewhat larger (though still mostly microscopic) and more complex animals. The bestiary includes plump, lumbering, eight-legged little critters called tardigrades, or waterbears, that waddle through the film of water around soil granules, searching for food. When they find plant roots or fungi, the "bears" drill into the cells with their sharp snouts, or stylets, and suck out the nourishing fluids. Another, more important group of plant eaters, the springtails spend their lives bumbling blindly through the dirt, munching on bacteria, fungi, dying roots, and the occasional dead waterbear or worm. Jumpy and alert, springtails are equipped to leap into action at the first tremor of danger. As their name suggests, they have spring-loaded tails that, when all is well, they carry tucked up under their abdomens. But when danger is detected, the tail fills with fluid and suddenly snaps down, sending the springtail catapulting to what, with luck, is safer ground. (The light-colored specks of "dust" that sometimes

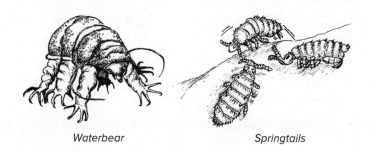

Waterbear   Springtails

somersault away from compost when the pile is turned are actually springtails taking flight from the disturbance.)

One square yard of prairie can support anywhere from 5,000 to 15,000 springtails in the top foot of soil. (That's about a square meter dug to the depth of 30 centimeters, enough to make a couple of wheelbarrow loads.) But these statistics pale compared with the totals for another major group of soil dwellers. The soil is crawling with microscopic mites, those scurrying, scuttling, eight-legged cousins of spiders and ticks. The same small patch of prairie soil that is hopping with springtails may simultaneously be home to up to 60,000 mites, representing dozens, or even hundreds, of species. (Although 48,000 kinds of mites have so far been identified around the world, the final catalog of species is expected to reach half a million or more.) While many mites make their living peaceably by feeding on microbes and plant roots, others are voracious predators—the tigers of this invisible world. *Spinibdella cronini*, for example, is a largish red mite (about the size of the

Predatory mite

period at the end of this sentence) that is common in prairie soils and that preys on soft-bodied mites and springtails. Its modus operandi is to seize its prey by surprise, tie it down with silk, and then suck the life out of it.

Pity the poor springtails. For "giant" mites are not the only menace in their shadowy world. They must also be prepared to flee from pseudoscorpions, jaunty little monsters that stalk through the soil with their venomous pincers armed and ready for action. Danger can also come swimming through the darkness in the form of carnivorous worms, innocuous-looking creatures (mostly smaller than the letter *i*) that either swallow their prey whole or attach themselves to the victim's body and scrape at it until the internal organs can be drawn out through the puncture. These silent predators are members of an enormous tribe of roundworms, or nematodes, that are the most numerous multicellular animals on the grasslands. Hundreds of species have been found at sites across the Great Plains, including those that feed on bacteria, protozoans, fungi, algae, mites, other nematodes, and the roots of higher plants, or some combination of the above. A heap of soil that contains thousands of springtails and mites typically also holds between three million and five million nematodes. Across the broad expanse of the grasslands, their numbers spin out of control, quickly mounting toward the astronomical.

GROWING SOIL

ALL SOIL IS full of life, whether under prairie, forest, tundra, or desert. Yet compared with other biomes, the temperate grasslands are remarkable for the extent to which they concentrate their resources in the underworld. The reason is simple: it's safer down there. The prairie climate is grueling— too hot, too cold, too wet, too dry, too wild and variable—but go a little way beneath the surface and the environment becomes surprisingly equable. During the growing season, for instance, when the thermometer can easily spike from 50°F to 105°F (10°C to 40°C), the temperature a few handwidths down in the soil remains pleasantly cool and virtually constant.

CHEMICAL WARFARE

Nematodes are not only the most numerous multicellular animals on the grasslands. They are also the most important herbivores. While aboveground plant eaters nibble on stems and leaves, herbivorous nematodes eat their way through the roots, attacking silently from beneath. One study conducted on a conservation reserve determined that the unassuming little worms worked their way through more plant matter per unit area than either bison or prairie dogs. And on pasturelands, the standing population of nematodes has been known to consume twice as much vegetation as a herd of cattle.

Fortunately, some plants are able to protect themselves against the attack of the killer worms. A number of common prairie forbs, including pasture sage (or fringed sagewort) and nodding onion, produce natural nematicides that ooze out of the roots and create a worm-free zone for themselves and their neighbors. Such protection is presumably one of the benefits of a diverse plant community in which the roots of different species are intertwined.

But not all plants are so community minded. Some use chemical warfare not to fight against pests but to resist the encroachment of other plants. For example, a dainty-looking short-grass species called prairie three-awn produces root exudates that kill nitrogen-fixing algae and bacteria. (These are soil organisms that capture nitrogen gas from the air and convert it into a water-soluble form in the soil, ready for plants to use.) Because prairie three-awn can get by without much nitrogen, this action creates ideal conditions for its own growth and keeps out competing species that need nitrogen-rich soil.

In a region where moisture is often the limiting factor for life, the soil also provides a buffer against drought. Long after the surface of the ground has been wicked dry by the sun and wind, the soil retains hidden reserves of moisture. Not only does water pool in the spaces, or pores, between the soil granules, but it is also glued to the surface of the particles by powerful chemical bonds. Thus, even apparently dry dirt often retains a thin film of water that is tightly bound to the soil grains. To tap this precious resource, grasses put down dense networks of thin, fibrous, probing, branching, fast-growing roots, all seeking moisture and nutrients. On average, grasses produce three or four times more roots by weight than they do leaves and stems, giving them a root-to-shoot ratio that is ten times as high as that of a forest.

The sheer volume of roots produced by prairie grasses is mind-boggling. In the mid-1930s, for example, one determined agrologist undertook to excavate and measure all the roots produced by a small patch of tall-grass prairie in Nebraska. From half a square yard, or meter, of ground, he extracted about 150 miles (240 kilometers) of fibrous roots—enough, laid end to end, to reach from Lincoln, Nebraska, to Sioux City, Iowa. The tally included 11 miles (18 kilometers) of green needlegrass roots, 13 miles (21 kilometers) of big bluestem, 23 miles (37 kilometers) of little bluestem, and 109 miles (175 kilometers) of Kentucky bluegrass. And that was without factoring in the silvery fuzz of absorbent root hairs that coat root fibers from base to tip. In another heroic experiment, a single rye plant (the domesticated cousin of several native prairie grasses) was found to have produced 7,000 miles (11,000 kilometers) of roots and root hairs in one four-month growing season.

In nature, these root systems obviously do not strike out across the continent. Instead, they squiggle and writhe into every nook and cranny of the soil, holding the dirt together with their slender, grasping fibers. Since most prairie grasses also produce lateral rootstocks, or rhizomes, that send down roots along their length, each plant—and each clump of root-bound earth—is connected to the next and the next. As plant intertwines with plant, and roots interweave with roots, the soil becomes tightly tied together in a thick, fibrous mat—the famous prairie sod, which the settlers used for building their first homes.

For thousands of years before the settlers arrived, that same deep fabric of roots had served as the foundation of the soil ecosystem. The enormous biomass of underground plant matter, both living and dead, provides an almost inexhaustible supply of food for soil organisms. Huge populations of herbivores graze on the roots themselves, including hordes of fungi, nematodes, and springtails. Still other microorganisms feed on the rich soup of nutrients that leak from the roots into the dirt. (Since these resources are concentrated in the rhizosphere—the soil that surrounds the root—much of the life in the soil is concentrated there as well.) In the dark underworld of the soil, everything gets eaten. The dead, the dying—even minute particles of excrement—are all on the menu for some miniature soil creature.

As a result of this endless round of digestion, organic material in the soil (principally dead roots) is first broken down and then gradually built back up into complex and relatively stable molecules known as humus. Dark brown or black, these humic substances not only give prairie soil its rich palette of tones but also contribute to its ability to support a dense tapestry of crops and native grasslands. To thrive, plants need to take in nutrients such as potassium and nitrogen (the K and N found in synthetic plant foods), which they soak up through their roots. Because these chemicals are water-soluble, they are readily leached down into the earth, beyond the depth at which the plants can reach after them. But if the nutrients are bound up in humus, they are effectively stored in the upper 6 to 12 inches (15 to 30 centimeters) of the soil. As the soil organisms eat away at the hard-to-digest humus, the stored nutrients are gradually supplied to the plants, much as if they were being freed from a time-release capsule.

Thus, prairie soils are fertile largely because—from the end of the Ice Age until the land was plowed—the native grasses have consistently produced more root fibers than the soil organisms have been able to consume. The surplus resources have collected in thick layers of dark, crumbly, humus-rich earth. By contrast, desert soils contain little or no humus, and forest soils have only a thin, upper horizon, or layer, of black dirt. Considered globally, intact and restored grasslands hold, or sequester, billions of tons of carbon, 50 percent more than is stored in all the forests in the world.

FORMATIVE INFLUENCES

TECHNICALLY SPEAKING, PRAIRIE soils are characterized by a deep surface layer, or A horizon, of topsoil; a lighter-colored B horizon of subsoil; and a band of calcium-rich salts somewhere within a couple of yards (about 2 meters) of the surface. But within the broad bounds of this definition,

THE TAXONOMY OF SOIL

Over the decades, North American soil scientists have devoted themselves to soil classification with such enthusiasm that they have produced two separate systems. The United States Department of Agriculture's taxonomy divides soils into eleven broad categories, or orders. Most grasslands soils are classified as mollisols (literally, "soft soils"), a designation that is subdivided by climatic conditions. Soils may be udic, ustic, xeric, or aridic as they progress from humid to parched; they step up from cryic to thermic and hyperthermic along the vector from frigid to warm. Other terms describe characteristics such as the texture and chemistry. Thus, a cool, dry soil from Montana is classified as an aridic cryoll (a dry, cold mollisol), while a warm, moist soil from Texas keys out as a thermic udoll (a warmer, wetter mollisol). Meanwhile, the rich farmlands of west-central Kansas are written down as fine, smectitic, mesic typic argiustolls, or silty, warm, well-watered mollisols with a heavy subsoil laced with slippery clays. And so it goes, in an outpouring of precisely defined vocabulary.

In Canada, scientists have retained an older system that is based more closely on observable properties. In this system, prairie soils are known as chernozems, from Russian for "black earth." Since chernozemic soils vary in color, this order is further subdivided into four major groups: Brown, Dark Brown, Black, and Dark Gray. Yet even this apparently straightforward system can show signs of linguistic strain. Is this a Calcareous Brown soil (light colored and high in calcium) or a Gleyed Solonetzic Dark Gray (somewhat darker, saline, and poorly drained)? Soil science is clearly not a profession for the tongue-tied. *See Map 7: Soils of the Great Plains.*

there is room for almost infinite variability. Soil is a unique and intimate blending of living and nonliving elements, and although grasses have doubtless been the major influence on the development of prairie soil, they have not acted alone. The physical setting of the Great Plains has exerted a powerful force of its own.

Just as the sea can be described as a body of water, so the soil could be defined as a body of weathered rock that, like the sea, is permeated by life. The nature of the soil ecosystem depends, at least to some extent, on the nonliving substrate, or "parent material," from which it developed. Grassland soils still bear the mark of events that happened thousands of years ago. For instance, the region immediately south and west of the Missouri River (from Montana and the Dakotas across to northwestern Nebraska) escaped the direct impact of the Ice Age. As a result, the grasslands there took root on an ancient, unglaciated landscape that had originally been shaped by the shallow seas of the dinosaurian age. Because of the chemical composition of those ancient deposits, the soils that formed on them are not particularly rich in some of the minerals that growing plants need.

Profile of a grassland soil, showing topsoil, A; subsoil, B; and parent material, C.

By contrast, soils to the north and east of the Missouri (from the Canadian prairies to northeastern Nebraska) have been built on mineral-rich gravels and ground-up rock left behind by the glaciers. And large parts of the central plains (from southeastern Wyoming, south into Texas and east to Kansas and Missouri) are blanketed in thick drifts of silt, or loess, that were deposited by fierce windstorms around the end of the Ice Age. Whether ice-borne or wind-borne, this glacial debris has provided the soil with a generous resource of plant nutrients that enhance its fertility.

In a number of places across the plains, the Ice Age also left behind a legacy of sand, a material that is much less favorable than silt for soil development. Sand is composed of relatively large particles, with large pores in between, a structure that permits water to run through it freely. As a result, organic material is flushed away before the soil organisms have a chance to set to work and very little humus is produced. In the millennia since the glaciers' last retreat, sand prairies like the Nebraska Sandhills and the Great Sandhills of Saskatchewan have produced only a fragile layer of topsoil that is susceptible to every passing footfall or gust of wind.

Yet the inability of sand to hold onto water also offers an unexpected advantage to certain plants. Because moisture is not tightly bound to the sand granules, it is readily available to the roots and often supports surprisingly lush stands of vegetation. The Nebraska Sandhills, for example, are home to a distinctive community of grasses and forbs, including specialists such as sand bluestem, sand dropseed, sand reedgrass, and the amazing blowout grass, which pushes its rhizomes out to a distance of 40 feet (12 meters), searching for bare sand in which to establish new plantlets. Similarly, the lemon scurfpea, or lance-leafed psoralea—a delicate-looking plant that is actually tough as nails—relies on a network of long, shallow roots that hold the soil in place and allow it to flourish even on the shifting sands of windswept banks.

The nature of sandy soils is largely determined by their geological origins. But most soils on the plains are even more strongly influenced by a second physical force—the climate. In particular, soil development is especially susceptible to regional differences in temperature and precipitation. In general, the more moisture that is available to grasses, the more roots they produce and the deeper and darker the topsoil that accumulates. Because precipitation increases along a west-to-east gradient across the Great Plains, soils follow a similar trend, increasing in fertility across the region. The thin, light-colored soils of the western steppe transition into the deeper, richer soils of the mixed grasslands to the north and east and culminate in the thick, black earth that developed under the tall-grass prairies.

Inevitably, in a world of intricate interactions, this pattern is complicated by a conflicting variable, namely, the north-to-south gradient in temperature. Where average soil temperatures are low, as they are on the northern plains, the soil organisms are sluggish and the rate of decomposition is slow, so humus accumulates in the topmost horizon. In the warmer south, by contrast, the more active microbes eat around the clock, causing rapid decomposition and slower rates of humus production. A glance at a generalized soil map *(See Map 7: Soils of the Great Plains)* confirms this effect. The rich black "mollisols" of the northernmost plains gradually give ground farther south to "alfisols" and "aridisols"—thinner, less fertile soils. At the same time, it is also possible to detect an overall improvement in soil quality from west to east, under the influence of increasing precipitation. (These gradients may become more marked in years to come, as high temperatures and low precipitation intensify in the southern plains, in response to human-caused climate change.)

MOVERS AND SHAKERS

IF THE CLIMATE has mediated the development of prairie soils over several thousand years, it also sets the pace of life in the here and now. Although the weather in the under-world is relatively stable compared with the chaos that reigns above ground, the soil ecosystem is not immune to climatic disasters. Drought is a particular challenge. Most soil organisms are aquatic species that have become adapted to live in the film of water surrounding the soil grains. If this habitat dries up (whether for a few weeks in the heat of mid-summer or during a prolonged drought), the microbes are thrown into a crisis. Although many individuals die, many more survive by encasing themselves in protective armor, transforming themselves into cysts, and becoming dor-mant. Waterbears, for example, shrivel up into little kernels called "tuns" that are capable of withstanding unimag-inable cold (approaching absolute zero), unspeakable heat (well above boiling), and brutal doses of X-rays (a thousand times the lethal dose for humans). When a piece of moss in a museum collection was moistened after 120 years, the long-dormant waterbears roused from their sleep and imme-diately started grazing.

Although waterbears may set the outer limits of biological endurance, most soil microorganisms have an extraordinary ability to "play dead" for months or years. A nematode egg can lie in the soil for up to a decade, waiting for its preferred plant host to appear. The minute that conditions improve—for example, when a dry crumb of soil is moistened by long-awaited rain—the soil organisms come back to life and start to eat and reproduce at a hectic rate. Life occurs in manic pulses—wet, on; dry, off—typically with peaks in the spring and autumn, and lulls in the hot summer months.

But there is one major group of soil organisms that has not adapted comfortably to the erratic moisture regime of the Great Plains. Earthworms, those most celebrated of all soil dwellers, breathe through their moistened skin, and for them, desiccation can be fatal. While most species can survive moderate drying by burrowing deep into the subsoil, coiling up in a knot, and entering dormancy, few can survive severe or prolonged water shortage. In the whole of the Great Plains, earthworms are naturally abundant only in the relatively well-watered soils of the east and southeast, including all or part of Nebraska, Iowa, Kansas, Oklahoma, Texas, and Missouri. In this region, the soil is home to worms of two dozen different species—gray, pink, brown, massive or delicate. Some are twice the length of a man's outstretched hand; others would fit neatly on a fingertip.

Among the most common earthworms in these soils are the so-called peregrines—hardy European and Asian species that have been transported to the prairies (and around the world) in the roots of potted plants. Ideally suited for life in agricultural soils, they have expanded their range tremendously in the last few hundred years, often at the expense of indigenous species. Of the six families of earthworms native to North America, four are known to occur naturally in tallgrass prairie soils. The populations currently seem to be strongest in relatively untouched regions such as the Flint Hills of Kansas, though even there the invasive, introduced species continue to intrude.

Nobody really knows what difference it makes when foreign earthworms appear on the scene, but it must have a profound effect on soil ecology. Earthworms are big eaters, and where they are abundant, they serve (in Aristotle's undying phrase) as "the intestines of the Earth," capable of producing their own weight in nutrient-rich castings every twenty-four hours. But not all earthworms are created equal, and experiments hint that native species may be specially attuned to meeting the nutritional requirements of native plants. When big bluestem was grown in pots with either the native earthworm *Diplocardia smithii* or the introduced pasture worm *Aporrectodea turgida*, the grasses produced significantly more roots in the presence of the homegrown species. Perhaps *D. smithii* squirmed in closer to the roots and delivered the nutrients where they were needed most. Or maybe the native worm produced hormonelike "growth factors" (as earthworms are known to do) that were particularly effective in stimulating the tissues of its longtime evolutionary companion.

In one way or another, the introduction of alien life-forms into tall-grass prairie soils must certainly be having complex repercussions. And the disturbance is presumably even greater in the northern and western plains, where earthworms appear to have been rare, or entirely absent, before agricultural settlement began. Across the northern span of the continent—to the southern limit of the most recent glacial advance—there are essentially no native species of earthworms. Either such creatures never existed or they all perished under the ice, never to be seen in modern times. (One telling exception is a small, pallid species called *Aporrectodea bowcrowensis* that was recently discovered in Alberta's Porcupine Hills, in an area that was not glaciated.) Native species of earthworms are also virtually absent from the short- and mixed-grass prairies of the western and central plains, a region that is chronically short of rain. The only earthworms to occur there are the immigrants that arrived with the settlers. Although little is known about their progress on this new frontier, they appear to be expanding into both grasslands and croplands, boldly going where no worm has gone in at least ten thousand years.

So, despite everything we've heard to the contrary, the presence of earthworms in the soil is not always a mark of good health. Across much of the Great Plains, it is a prime indicator of disturbance. In arid regions, the work of digesting organic matter is assigned by nature not to tender worms but to the invincible legions of drought-hardy microorganisms. Meanwhile, an unlikely consortium of animals has taken on some of the earthworm's other tasks. For instance, the job of turning the soil—digging in dead vegetation from the surface and bringing up clay from below (a service that is provided by earthworms where they are plentiful)—is here performed by burrowing mammals, including ground squirrels, pocket gophers, prairie dogs, and badgers, and most notably by ants. Given their impact on prairie soil, it could almost be said that ants are the earthworms of the Great Plains Grasslands.

There are dozens of different species of ants on the prairies (eighty-three, for example, in North Dakota alone), and while they differ in many particulars—size, color, diet, habitat—they tend to have one characteristic in common. Unlike ant species elsewhere that may take shelter in rotten logs or inside rolled-up leaves, most of the ants on the grasslands nest in the soil. An anthill is merely the visible extension of a complex of tunnels that extends down through

AN ASTONISHMENT OF ANTS

The fascination of ants extends far beyond their importance in building the soil. Although prairie ants represent only a tiny fraction of the fifteen thousand species that have been described around the world, they nonetheless exhibit a full range of weird and wonderful behaviors.

By and large, prairie ants are carnivores with a taste for sweets. Because they hunt in groups, they are generally the most successful predators in their size range, subduing impressive numbers of spiders and smooth- and soft-bodied insects. Prey animals are generally torn into bits and carried back to the nest, to be fed to the larvae. The worker ants' main source of nourishment, however, is nectar from plants. Many plants, including native sunflowers and partridge peas, have glands on their leaves, stems, or buds that produce ant-sized drops of nectar. Ants sup at these honeyed fountains— "extrafloral nectaries"—and defend them vigorously, in the process inadvertently defending their hosts against plant-eating insects.

Formica incerta

Ants also acquire nectar at secondhand, by milking "honeydew" from the anuses of sap-eating aphids. Some ants carry on this trade in public, on the leaves of aphid-infested plants, but others prefer to milk their aphids in privacy. Except for nuptial flights—when winged "reproductives" burst forth to mate and form new colonies—these ants spend their entire life cycle underground. Their nests feature special chambers around the roots of selected plants, where they keep their "herds" of root-feeding aphids. The ants tend their livestock constantly, tapping the aphids with their antennae to induce them to secrete sweets and slaughtering the tender young to supply their need for meat.

Lasius neoniger

the topsoil and into the subsoil beneath, often to a depth of 3 to 6 feet (1 to 2 meters). The labor of excavating these underground runways and lugging the soil up, grain by grain, can occupy a colony of several thousand ants for the better part of a decade.

But this Herculean effort is justified by the benefits that accrue to the ants. Like all insects, ants are ectothermic, or cold-blooded. In other words, they assume the temperature of their surroundings and regulate their body heat by seeking shade or sunlight, as required. By constructing their complex, multilevel dwellings, ants alter their environment to create a range of microhabitats, each with its own particular microclimate. In winter, for example, they may retreat to their deepest chambers (below the reach of the frost), where they shuffle groggily through the coldest months. In summer, some species make sophisticated use of their anthills, which are carefully angled to soak up rays from the sun. As heat builds up in the hill, crews of workers lug their larval brood from place to place inside, striving to keep both themselves and their charges at just the right temperature.

As ants manipulate their surroundings to meet their own needs, they inadvertently replenish the soil. The particles of clay that they carry up from their basements are rich in the minerals needed by plants and thus help to maintain the fertility of the upper layers. One estimate suggests that the top 2 feet (60 centimeters) of prairie soil is turned over every hundred years through the infinite efforts of ants and other soil creatures. And all this is achieved without any attention from us, beyond the farthest horizons of our senses.

CHAPTER 5

HOME ON THE RANGE

We did not think of the great open plains,
the beautiful rolling hills,
and winding streams with tangled growth as wild…
Earth was beautiful
and we were surrounded with the blessings of the Great Mystery.
LUTHER STANDING BEAR, OTA K'TE,
MY PEOPLE THE SIOUX, **1928**

F I HAD to name my favorite place on the prairies, it would be the high, arid benchlands that sweep along the rim of the Frenchman Valley in southern Saskatchewan. Climb up there on a blue day in early September, out and onto the bald steppe at the top, and the wind will slam against you as if it had a grudge against anything that dared to raise its head above the grasses. This is a landscape that has held to the horizontal for thousands of years, even resisting the torrents of meltwater that once rushed down from glaciers in the Cypress Hills and gouged out the wide, flat trench of the river valley. From up on the benches, you can see the descendant of this ancient flood, a soapy, sleepy little stream that writhes through its oversized course, as if trying to make up in complexity for what it has lost in force.

High on the benchlands, the grasses are stunted and crisp, and the ground bristles with clumps of prickly pear cactus. At one time, long before the Ice Age, this drought-stricken upland was itself the course of a great river that flowed down from the young Rocky Mountains across the northern plains, burdened with loads of gravel and debris. As stones were dropped by the river, they settled onto the riverbed, where they formed a thick layer of sediment. (As the ages passed, this pavement helped the land resist erosion by water and wind, with the result that the former valley is now a high tableland.) Millions of years later, the benches of the Frenchman Valley are still strewn with rounded, river-washed stones, most of them two or three times the size of a clenched fist, all speckled and splotched with lichens.

The eye scans the stone-pocked surface, searching for clues. Grass, sky, rocks. Scattered rocks that suddenly swerve to follow a curve, a curve that coalesces into a circle. Two concentric circles of stones lie half-buried in the grass, tracing a circumference of about twenty paces. A few steps away, another ring and another. People have been here before us in the fairly recent past, though whether it was two hundred years ago or two thousand is more than we can tell. All we know is that a party of bison hunters once chose this bank of the valley to make their camp and that they used these rings of rounded stones to anchor their skin tents. Downslope from the tipi rings, where the benchland begins to fold toward the valley, smaller rocks and pebbles have been washed into

INTO THIS WORLD OF BEAUTY

When a Kiowa woman named Old Lady Horse (Spear Woman) looked back on the past, she recalled the not-so-distant time when all the necessities of life had been provided to her people by the bison. Hides for shelter and clothing, bones for tools, blood and meat for food. "The buffalo were the life of the Kiowa," she said.

When Europeans came to the plains to build railroads and raise cattle, the bison did their best to protect the Kiowa from harm. "They tore up railroad tracks and the gardens," Old Lady Horse recollected. "They chased the cattle off the ranges. The buffalo loved their people as much as the Kiowas loved them."

But when the newcomers sent in soldiers and hide hunters, the buffalo admitted defeat and gathered in council to decide what to do. As it happened, the Kiowa were camped on the north side of Mount Scott, Oklahoma, at this time. Early one morning, a young Kiowa woman looked up from camp toward the mountain, through the mists over Medicine Creek, and saw the last bison herd appear like a spirit dream. As she watched, the face of the mountain opened and the bison walked inside, into a world of plum blossoms and freshness, where "the rivers ran clear, not red." Into this world of beauty the buffalo walked, and the mountain closed behind them and they were gone.

Old Lady Horse told this story to Alice Marriott, who included it in her book *American Indian Mythology* in 1968.

terraced drifts, and some of them are chipped and sharpened along one edge as if they had been fashioned for working skins. They fit neatly into the hand.

This land has never been disturbed by the plow—not even a stone has been touched—and it would be easy to think of it as wild, the last stand of the great North American grasslands. In the beat of the wind, you can almost believe you are hearing the muffled drumming of a bison herd that, any minute now, will come rolling into view over the horizon. Through bones and stones and life-forms, these lands conserve not only the memory of the past but the whole promise of a future for many grassland animals and plants. The surviving native grasslands bring us as close as we can now get to the prairie in its natural abundance. Yet when you walk toward that horizon and peer down into the valley beyond, you will not find wild herds or hunting encampments. Instead, you are very likely to see a bunch of cows. It's still fabulous out there on the rangelands, but it ain't exactly wild.

THE INTERCONTINENTAL BOVID BOOGIE

RANGELANDS—EXPANSES OF native grassland that are grazed by livestock—exist only where the prairie has somehow managed to escape the plow, usually because the soil is too dry, too thin, too rocky, or too steep to be suitable for crops. The greater the agricultural potential of a region, the less native prairie is left. Because moisture and soil fertility improve from west to east and from south to north, the percentage of the land in natural cover increases in reverse, from east to west and from north to south. On the short grasslands of the southwestern plains, where the rains are meager and the soil is relatively poor, something on the order of 70 percent of the landscape has survived with its natural integument of grasses and wildflowers. A little to the north and east, by contrast, in the slightly moister mixed-grass zone, the percentages drop off, and what little prairie survives is badly fragmented. (In Manitoba, to cite the most extreme example, less than 0.1 percent of the original mixed-grass landscape remains intact.) On the whole, the irresistibly fertile tall-grass zone has fared worst of all, with less than 1 percent of the entire ecoregion remaining in a more-or-less natural condition. This state of affairs earns the tall-grass prairie dubious honors as one of the most ravaged eco-regions in North America, officially classified as Critically Endangered. Today, the only large, continuous blocks of tall grasslands lie in the Flint Hills of eastern Kansas and the Osage Hills of northeastern Oklahoma, where a stony rime of crystalline quartz just beneath the surface long ago put a dent in the enthusiasm of plowmen. *See Map 8: Grassland Cover on the Great Plains.*

The surviving native grasslands span the complete spec-trum of prairie types, from the stony benchlands of the Frenchman Valley and the rollicking dunes of the Nebraska Sandhills to the sculpted badlands of the Palo Duro and beyond. Ecologically, they are as different from one another as big bluestem is from blue grama or as a sage grouse (a threatened bird of arid prairie) is from a prairie chicken (its threatened cousin on the humid tall grasses). Yet for the last century or so, these diverse grasslands have all answered to the same set of human demands—to produce food for large numbers of cattle and ultimately for ourselves.

In some ways, the introduction of domesticated livestock onto the Great Plains was not much of a shock to the ecosystem. Bison and cattle belong to the same evolutionary lineage, the family Bovidae, and trace their ancestry back to India and China some two million years ago. But whereas the ancestral bison headed north across the Bering land bridge into the Americas, cattle took a turn to the west and wandered through the Near East and into Europe. (They are thought to have been domesticated in or around what is now Turkey some ten thousand years ago.) True to their common ancestry, the two species filled much the same ecological role on their respective continents, as large mammalian herbivores that specialize in consuming grasses and, to a lesser extent, other forbs, or nonwoody plants.

Since grasses are ribbed with particles of silica (grains of glassy sand), both bison and cattle have self-renewing teeth that grow in to compensate for wear and tear. And since grasses and other plants are composed mostly of cellulose, which mammals cannot easily digest, both species have fermentation vats for stomachs—elaborate four-chamber structures in which a frothing stew of microorganisms breaks down the tough plant fibers. Halfway through this process (between stomachs two and three), the animals cough up a "cud" of half-digested mush, give it a thoughtful second chew, and send it back down for another round of microbial demolition. By collaborating with these small-but-mighty microbes, large ruminants like cattle and bison free up nutrients locked inside plants and keep resources cycling through the ecosystem.

As they convert plant tissues into cold-cuts-on-the-hoof, herbivores inadvertently provide a source of highly digestible food for the next trophic level in the system, the carnivores. The carnivores, in their turn, inadvertently assist the herbivores in avoiding predation. By removing the halt, lame, and unwary from the breeding stock, predators select for individuals that are best adapted to escape their grasp and hence make successive generations of prey more difficult to catch. Thanks to this rigorous breeding program, bison and cattle have both acquired large eyes and excellent vision, the better to help them detect predators. They have become strong and sure-footed, able to sprint away from danger or to travel long distances to seek safety. Most important of all, both species have become overwhelmingly sociable. Like many other animals that live on the plains, bison and cattle find comfort in the company of others of their species. Not only can they benefit from the combined vigilance of the group—all eyes peeled for a possible threat—but they can also attempt to use their neighbors as "bovine shields." The more potential victims that are available to a predator, the less chance there is that any given individual will end up as a sacrifice to trophic exchange.

So even though they spent the last half million years or so on opposite sides of the Atlantic, bison and cattle are fundamentally alike. Removing wild American bison and replacing them with tame Eurasian cattle—though a stunning act of hubris—was ecologically relatively neutral, the substitution of one large, ruminant herbivore for another on a landscape that had sustained large herds of grazers since the retreat of the glaciers. (The same cannot be said for the desert and semidesert grasslands west of the mountains, which had not evolved under heavy grazing pressure from bison and which have suffered serious damage since livestock were introduced.) The ease of this transition is demonstrated by the suite of animals that once lived in, under, and around bison pats. As nearly as anyone can tell, this group, which includes several dozen species of dung beetles and several specialized mites among many others, took to their new diet of cow pies without hesitation. And the species that dine on the dung dwellers—a diverse group that includes blackbirds, small rodents, and (in the tall-grass prairie) the beautiful little ornate box turtle—also seem to have made the switch without any noticeable problems.

Ornate box turtle

HOW GRAZING SHAPES GRASSLANDS

THIS IS NOT to say that the introduction of cattle to the Great Plains has been completely benign. It has not. Cattle and bison, though similar, are not identical. A case in point: bison like to throw themselves on the ground and flail around in the dirt, a self-care routine that is thought to coat the skin with dust and offer protection from biting insects. In the process, they wear away shallow bowls, or "wallows," in the earth. By rubbing out the grasses from these hollows, bison create openings for other kinds of plants, often moisture-loving species such as prairie dropseed, together with various rushes and sedges. Because they increase the diversity of plants that are available for shelter and food, bison wallows also augment the diversity of insects, birds, and mammals that the local landscape can support. And if the depressions fill with water (as they may do in the spring), they provide seasonal habitat for aquatic insects and water-loving shore-birds. Or at least they used to.

Because cattle do not wallow, this dynamic has been lost. Meanwhile, cattle have pushed the ecosystem in new directions. Less well adapted than bison to extremes of temperature and drought, cows spend more of their time in the shelter of trees and around sloughs and watercourses. This behavior severely affects both water quality and the condition of riparian, or bankside, forests. What's more, although cattle and bison are both grazers with a special fondness for grass, there are subtle but significant differences in their diets. Bison prefer a steady diet of grasses, with just a garnish of other plants (forbs and shrubs make up only 5 to 10 percent of their diet). Cattle, by contrast, choose grass as their staple food, with a side salad of mixed greens (forbs and shrubs provide between 20 and 40 percent of their feed). Although no one knows how the vegetation of the rangelands has changed since the bison were displaced—which grasses and forbs have become more common and which more scarce—Bossy the Cow's exotic dietary preferences must have had a significant impact.

Herbivory shapes grasslands. The first principle of range management is that grazing causes shifts in the relative abundance of the many and varied species of grassland plants. In mixed-grass prairie, for example, blue grama, buffalo grass, and pasture sage tend to become more common on range that is grazed by cows; these plants fall into the class known as increasers. Conversely, midheight grasses such as needle-and-thread and western wheatgrass become less abundant and hence are called decreasers. (It has even been suggested that the short-grass prairie, with its mat of increasers like blue grama and buffalo grass, is really just mixed-grass prairie that has been subjected to heavy grazing.) Some plants—like the big, bold butterfly milkweed of the tall-grass prairies—vanish from the range the second they appear because the cows enjoy eating them. Out in the pasture, grizzled rangemen shove their Stetsons back off their brows and lean against their pickups to discuss the status of these "ice-cream plants" in their pastures.

Until fairly recently, range managers have worked on the comforting assumption that the disruptions caused by grazing could be corrected simply by removing the cattle and letting the pasture rest. This belief was based on what seemed at the time to be solid science: the cogent theories of a pioneering ecologist from the University of Nebraska, Frederic E. Clements. As Clements explained it in the 1910s and 1920s, mature, natural grasslands were the endpoint, or climax, of a linear process of plant succession. This climax community, he said, represented a state of equilibrium in which the conflicting needs of the various plant species (their demands for nutrients, water, and light) were held in such fine balance that the grasslands, if left to themselves, would wave on unchanged forever. But any disturbance, such as heavy grazing, fire, or drought, upset this steady state and sent the system tumbling back into an immature and unstable stage. Cattle, for example, not only evicted the ice-cream plants and decreasers from the climax community but in so doing opened the prairie to early successional species such as broomweed and Russian thistle. These same "weedy," or invasive, plants were always first on the scene when plowed lands were abandoned or pastures were degraded by overgrazing.

The good news, as Clements saw it, was that plant succession was a predictable and reversible process, in which pioneering species inevitably created conditions for their successors, which created conditions for their successors, and so on, until the climax community was ultimately restored. Thus, any damage caused to the rangelands by overgrazing could be repaired, in due course, through the healing processes of nature. If only this were true. Instead, it turns out that succession is more fickle than the good professor believed—more subject to the vagaries of climate and simple luck. Once the prairie has been disturbed, there is no great likelihood that it will grow back in, plant for plant, as it was before it was disrupted. It is even possible that it will not come back at all. In the last hundred years, for example, large areas of the southwestern United States have been permanently converted from grassland into a prickly shrubland of mesquite and creosote bush, probably as a result of heavy grazing during the 1800s and 1900s.

DON'T FENCE ME IN

BUT THE MAIN ecological challenge on the rangelands—the real thorn in the side of the ecosystem, so to speak—is not the absence of bison or the presence of cattle. It is barbed-wire fencing. The "devil's rope," as it was once known, went into mass production in the mid-nineteenth century to meet the needs of settlers moving onto the treeless (post-and-rail-less) plains. By the early 1880s, factories in Illinois were spinning out enough barbed wire every year to make twenty-five complete twists around the equator. Not surprisingly, the Great Plains were quickly crisscrossed with fences, and by the close of the decade, the era of the open range had ended. With it went the free interaction between grazers and grasslands.

Before fences, there had been movement. Every spring for millennia, the bison had poured out of their winter ranges on the foothills and parklands, massed together in vast herds, and flowed across the continent on a quest for greener pastures. Like the barren-ground caribou of the tundra, they had charted new routes each year, jogging east to track recent rainstorms, north to follow the spring, south to savor the green-up after a fire. But wherever their travels took them, they always looped west again in the fall and followed a halting, zigzagging course back toward their wintering grounds. There, in the dead of winter (when the tall grasses of the eastern plains had little value as food), the bison sustained themselves on a steady diet of sun-cured, protein-rich blue grama and buffalo grass. Thus, season by season, the bison tracked the resources of the plains and responded to the interactions of the climate, the soil, and the plants.

What happened wasn't always pretty. Even with the freedom to move at will, tens of millions of bison couldn't always find enough to eat. In fact, the fossil record suggests that the species disappeared from the plains entirely for the better part of 3,000 years, between about 4,500 and 7,500 years ago, when the climate was much drier and warmer than it is at present. More recently, during the ferocious winter in 1876, several million bison starved to death along the Brazos River

ALIEN INVADERS

Since the mid-1800s, hundreds of novel plant species have been introduced into the grasslands of North America. For example, Eurasian grasses such as crested wheatgrass, smooth brome, Kentucky bluegrass, and tall fescue were initially planted as forage for cattle or (during the drought of the 1930s) to protect the soil from erosion. Since then, however, these exotic species have advanced beyond all bounds, outcompeting the native species and crowding them out.

Crested wheatgrass

Happily, these intrusions are not always a dead loss. To a herbivore looking for fodder or a songbird building a nest, grass is a useful resource, whatever its evolutionary origins. So while these plants arrived as intruders, many of them make a contribution ecologically. But there are exceptions. Leafy spurge, for example, is thought to have landed from Europe in 1827, in the ballast of a sailing vessel. Freed from European diseases and insects, the newcomer laid claim to the continent, infesting both tilled fields and overgrazed pastures. Today, leafy spurge continues to advance, especially across the northern rangelands. Not only does it outcompete native plants for nutrients, water, and light, it also exudes toxins from its roots that prevent other plants from putting up shoots. What's more, it is defended against herbivores by poisons in its leaves, which are repellent to cattle, horses, and many wild species. A few types of European flea beetles and caterpillars have been introduced to limit the plant's spread, but even they cannot eradicate it. Resistant to every known method of control, leafy spurge—and other plants of the same ilk—are a menace to grassland ecosystems.

of southern Texas, so many that their humped carcasses reminded one observer of a field of pumpkins.

But if bison sometimes felt the sting of random misfortune, they could also dish it out. After a herd of bison had trundled across the prairie, the prairie generally looked as if a herd of bison had trundled across it. (People traveling in the wake of a herd often complained that they couldn't find so much as a blade of grass for their horses.) Fortunately, grasses are uniquely equipped to put up with this kind of abuse. Unlike shrubs and forbs, which produce new growth from buds on their branches and stems, grasses shoot up from buds at (or even slightly under) ground level. This means that if a grass is eaten, it can immediately send up fresh leaves, whereas a shrub or a forb may be done for until the following spring. What's more, grasses that have been defoliated often shift hormones and other resources from their roots to their new shoots, so that their rate of photosynthesis increases dramatically. The loss of leaves to a herbivore causes some prairie grasses to grow so exuberantly that they slightly overcompensate for the tissues that have been removed. It probably doesn't hurt that the saliva of some grass-eating mammals contains growth stimulants that encourage the grasses to send out new shoots.

Not all grasses are equally well equipped to cope with the stress of being chewed. As a rule, increasers, a.k.a. grazing-tolerant grasses, have a greater capacity to fight back than do decreasers. For example, crested wheatgrass—an introduced species from Asia that can be recognized by its handsome fishbone seed heads—puts out replacement leaves two or three times faster than does its native relative, bluebunch wheatgrass. This competitive advantage helps explain why crested wheatgrass has been able to crowd out native plants and establish itself across the plains at their expense.

To get the true measure of the relationship between grasses and grass-eating mammals, it helps to see the natural dynamics that evolved between them. Today, this is only possible in the few places on Earth where migratory herbivores are still free to roam, such as the tropical grasslands of the Serengeti Plains in East Africa and the temperate grasslands of Yellowstone National Park in Wyoming. Despite the obvious differences between these two ecosystems, researchers have discovered a surprising similarity between them. In both places, grasses that are grazed, each in their season, by migratory herds produce far more forage than do grasses that are protected from grazing (43 percent more in Yellowstone, 70 percent more in the Serengeti, according to one report). This bonus far outstrips any "overcompensation" that occurs when animals are confined within fenced pastures.

The unexpected synergy between migratory herbivores and grasses is probably the result of a chain reaction that begins when the leaves are cropped. The grass plant responds to this insult by shedding organic molecules from its roots. These molecules provide food for soil microbes, which go into overdrive processing humus and releasing nutrients in a form that plants can use. Thanks to this input of fertilizer, the plants experience vigorous growth, and the range is superabundantly replenished. But this response only appears to happen if the herbivores are free to migrate and select grasses in the juicy prime of youth. If the stress of herbivory is delivered in a well-timed pulse, the plants may actually benefit from it.

Did the bison and the prairie grasses once engage in a similar interaction? No one knows for sure. But the loss of this cyclical rhythm may be the price we pay for curbing the wild prairie with barbed-wire fencing.

PROMOTING PATCHINESS

ON THE OPEN range, bison (and cattle, for that matter) had gone wherever they chose and moved on whenever they liked. If they traveled through an area quickly, snatching mouthfuls on the run, the grasses were left standing in tattered clumps. If they lingered, the prairie was reduced to a worn-out, close-cropped lawn, where whatever remained uneaten was trampled or splattered with dung. These impacts occurred haphazardly, tracking the uncertainties of rain, lightning, and fire, to say nothing of the whimsies of the bovine mind. Once a herd had passed through an area, it might return in six months or five years or never again. As a result, the Great Plains were an ever-shifting mosaic, in which a patch of tall, ungrazed vegetation might stand next to a mouthful that had been cropped to the quick. On a larger scale, broad tracts of land might lie untouched (too dry to attract attention), while the green-up after a sweeping prairie fire attracted large herds of bison. Thus, at their most vibrant, the grasslands were a kind of living crazy quilt, with patches of vegetation of varying sizes in varying states of recovery from grazing, fire, or drought. This is the true climax condition of the Great Plains Grasslands, a system that achieves stability by responding constructively to continual challenges.

The patchiness created by grazing and other disturbances provides a diversity of habitats for a diversity of prairie residents. (Under present-day conditions, the rotation of cattle from one fenced pasture to the next can be used to simulate this richness and complexity.) Take birds, for example. Of the 260-odd species of birds that regularly breed on the Great Plains, most are species that moved onto the grasslands from other (mostly forested) parts of the continent at the end of the Ice Age. By and large, these immigrant species still cling to the traditions of their ancestral homelands and nest in trees and shrubs. But a small, select group of species—a couple of dozen in all—prefer the open plains, where they nest in, or under, the forbs and grasses.

Within this Lilliputian jungle, the prairie-adapted species seek out the special conditions that each needs for its nests. Some, like the bobolink—a flashy black-and-white blackbird that enlivens the mixed- and tall-grass prairies of the north-central and southern plains with its bubbling song—require dense and expansive stands of thigh-high flowers and grass. The female typically chooses a site that is shaded from the grilling sun (often at the base of a large forb, such as meadow-rue) but where the canopy is open enough to let light filter through. Here, in the cool, dappled shade, she scratches out a patch of bare earth and constructs her finely lined, cup-shaped basket. By contrast, the grasshopper sparrow—a widely distributed but inconspicuous "little brown job" with a shrill, insectlike buzz—opts for more open habitats, with scattered clumps of tangled, mixed-height plants. Its nest is buried in a thicket of vegetation and domed over with grass, an object lesson in the arts of concealment.

Each species looks at the grasslands from its own unique point of view, seeking patches of vegetation that are the right height, density, and structure to meet its needs. Thus, although the grasshopper sparrow and the maddeningly similar Savannah sparrow both are found in open habitats, the former chooses sites with more patches of bare dirt (for catching insects), the latter those with less. And so it goes. The chestnut-collared longspur—so called for the rich splotch of color on the neck of the breeding male—seeks out closely cropped rangelands with the odd rose bush or tuft of grass for nesting cover. Meanwhile, its drabber but doughtier cousin, the McCown's longspur, selects only the most barren prairie, where it nests in the shade of a prickly pear cactus or a cow pat. And the horned lark—another, quintessential short-grass species, much loved for its sweet, slurred song—prefers patches that have been thoroughly trampled, where it can hunker down in the dirt with little more than a stone or a tuft of grass for shelter.

Trying to remember the particulars of which species needs what and why taxes the memory. But for millennia, the turbulent dynamics of the prairies have provided for all these conflicting needs without any conscious intention or evidence of strain. Habitat simply happened. For all its chaotic uncertainties, the prairie ecosystem generated enough stability to sustain even those species with the most complex and demanding needs. The upland sandpiper, for example—a jaunty member of the shorebird tribe that has adapted to life on the arid plains—requires short vegetation with visible perches while courting, knee-high growth with a moderately open structure for nesting, and closely grazed pastures for feeding and rearing its chicks. What's more, these resources must be provided generously enough to support not only a single pair but a loose colony of breeding birds. The sight of an upland sandpiper perched on a fence post is mute proof that these demands have been met in one way or another for thousands of years.

Today, however, two of the main forces that drive the prairie ecosystem have been muted. Wildfire has been suppressed. (On the tall-grass rangelands of Kansas and Oklahoma, annual burns still occur, but they are generally timed to promote the growth of grass and to reduce the diversity of "weeds" and other flowering plants.) In addition, the migratory ebb and flow of herbivores across the plains has been halted. Fortunately, even when confined by fences, cattle help to maintain patches of vegetation—clipped versus tall, open versus dense—and this effect can be enhanced by implementing an appropriate regime of management. By manipulating variables such as stocking level (number of cattle), season and duration of grazing, and rotations of grazing and rest, ranchers can manage the prairie to provide an array of habitats. The best and wisest land managers do exactly that because they understand that rangelands with a natural diversity of vegetation will outproduce and outlast those that are reduced to homogeneous spans of grass. But even with the most beneficent intentions, it would be a tall order to manage grasslands for commercial cattle production and, at the same time, to maintain the prairie in its free flow of creative change.

There is no doubt that the populations of many grassland birds have declined in the last two hundred years. There is no doubt at all that those declines are continuing today. Data from the Breeding Bird Survey (a standardized count of birds

HOOT, BOOM, STRUT, DANCE!

Four species of grouse are native to the Great Plains: the sharp-tailed and greater sage grouse, and the greater and lesser prairie chickens. Despite the differences in their common names, they are so closely related that, where their ranges overlap, they occasionally interbreed and produce fertile hybrids. *See Maps 9A and 9B: Prairie Chicken/Grouse Distribution.*

Although sharp-tails are still common throughout much of their range, their southernmost populations have vanished in the last hundred years, the combined result of overhunting and habitat loss. Sage grouse

Greater prairie chicken

and prairie chickens persist as scattered and declining populations on fragmented habitat, and their long-term survival is in question. But who could bear to lose them, if only because of their antic, exotic, ecstatic performances? In the cold dawn of early spring mornings, the male grouse of each species gather in groups to compete with each other for status and access to females. First, they establish themselves on an open piece of prairie (stubble or closely grazed grass), where each bird stakes out a miniature territory. Within this ritual domain, he faces off against his neighbors in a twirling, gobbling dance, puffing up pouches on his neck or chest to maximum effect. The object of these strange proceedings is to claim a territory at the center of the dancing ground, or lek, and thereby impress the hens as a top-quality prospect. The dominant male in the lek fathers most of the chicks and makes his contribution to the survival of the species.

that has been conducted annually since 1966) documents recent declines in thirty-two of thirty-seven grassland species. For many of them—including the bobolink, horned lark, and grasshopper, Savannah, and vesper sparrows, together with other favorite prairie species such as longspurs and meadowlarks—the losses are statistically significant. This finding removes any hope that the downturns are merely random blips on the graph. Overall, grassland birds have suffered steeper, more consistent, and more geographically widespread declines than any other group of birds in Canada or the United States. These losses are caused in large part by the relentless conversion of native prairies to croplands and housing estates. The less native grassland, the fewer grassland birds. But these doleful statistics also help underline our failure, as a society, to reward ranchers for managing their land for "patchiness," or biodiversity, and to pay them for producing bobolinks as well as T-bone steaks.

FOOD FOR SEX

ALTHOUGH CATTLE ARE the most conspicuous herbivores on the range, they are certainly not alone. Ounce for ounce or gram for gram, natural grasslands support a greater live weight of plant eaters than any other terrestrial ecosystem. (And this statement does not take into account the zillions of nematodes and other soil animals that dine on the plants from below.) The largest group of grassland herbivores, in numbers if not in mass, is insects. While a complete catalog of prairie insects has not yet been compiled, the list of the herbivorous species alone would certainly contain several thousand entries, perhaps even tens of thousands. There are insects that feed on plants by piercing through the cell walls and sucking juices out (aphids and "true bugs"); others that rasp a hole in the plant to obtain nourishing fluids (thrips); still others that are equipped to feed on nectar by lapping (bees), sponging (certain flies), or siphoning it up with their tubular tongues (butterflies and moths). Several groups have powerful jaws that permit them to gnaw on seeds, like the worker caste of harvester ants, or to feed on leaves and stems, like the hungry legions of sawflies, beetles, grasshoppers, and caterpillars. With their stabbing beaks, hairy tongues, jutting mandibles, and other specialized mouthparts, these insects look as strange, to human eyes, as the weirdest creations of science fiction.

Yet despite their otherworldly appearance, plant-eating insects are of down-to-earth importance in grassland ecosystems. On the one hand, like all herbivores, they prune and shape the vegetation, one miniature mouthful at a time, thereby contributing to the patchiness of the prairie. On the other hand, they inadvertently provide a generous source of convenient, bite-sized meals for a wide array of small-to-medium-sized meat eaters, including predacious beetles, ants, spiders, songbirds (even seed eaters start their young on an all-arthropod diet), kestrels and other small raptors, and a few oddball rodents—like the thirteen-lined ground squirrel and the northern grasshopper mouse—that are highly insectivorous. But the critical ecological function of plant-eating insects is also the most riveting. They often serve their food plants as sexual go-betweens.

To produce seeds, flowering plants must somehow transfer pollen from the male organ, or anther, to the top of the female organ, or stigma. While some flowers (including the closed, late-season blossoms of wild violets and narrow-leaved puccoons) can accept their own pollen and are thus self-pollinating, most must receive sperm cells from other members of their species. Unable to go a-courtin', some of these cross-pollinated plants rely on the wind to distribute pollen for them. Self-pollination and wind-pollination are both common among grasses. Most wildflowers, by contrast, rely on insects (including beetles, beeflies, butterflies, moths, wasps, and bees) to carry their pollen from plant to plant in a transaction that could be summarized as food for sex. Insects that are enticed into flowers for a feed of nectar pick up sticky pollen grains, some of which they deposit in the next blossom that draws them in for a taste.

Although many nectar-and-pollen-feeding insects are generalists that visit a variety of different plants, some have very specific preferences. For example, of the two hundred species of pollinating bees that are found on Missouri prairies, most are widely distributed in other habitats. But a significant proportion of the total—about 25 percent—occur only on native grasslands. Their allegiances are so specific that of the prairie specialists, twenty-four species work only the uplands, thirteen species are restricted to sand prairies, and another ten species are found exclusively on bottomlands. Different bees for different plants. Sometimes these alliances are easy to observe. In Missouri, for example, the handsome black-and-yellow bumblebee *Bombus nevadensis* is seen almost exclusively on the heads of wild bergamot, where it sends its long tongue into the equally long tubular corollas of the blossoms. By contrast, the all-yellow *Bombus griseocollis*, though more of a generalist, tends to favor plants like purple coneflower and other species of *Echinacea* that hold their nectar within reach of its midlength proboscis. A third species, *Bombus impatiens*, is attracted to the delicate fringe of blooms along the spike of Culver's root, which can be penetrated by its short tongue. These teams of plants and insects have evolved together to meet the needs of both, providing the insects with a reliable (and sometimes almost exclusive) source of food, and the plants with an eager crew of sex workers. Through this symmetry of behavior and body parts, they have kept the grasslands buzzing and blooming for millennia. By protecting and restoring natural prairies, and enriching tame pastures with flowering plants, we can help to sustain these life-giving relationships.

A PRAIRIE DOG HOME COMPANION

AFTER INSECTS, THE second-most populous group of herbivores on the grasslands is rodents. Although all the members of this tribe are equipped for a life of gnawing with a pair of big, self-replacing front teeth, different species have become specialized to gnaw on different types of plants. Pocket gophers, for example, are stubby, squirrel-sized mammals with small, weak eyes, that spend most of their lives beneath the surface of the soil. There, they subsist mainly on the roots of perennial forbs, such as yarrow and goldenrod, which they locate by tunneling through the ground. As they "swim" through the dirt, they first kick the soil behind them, then turn around and, using their flat foreheads, shove it back along the burrow and onto the surface. These mounds of loose soil, as tousled as unmade beds, provide openings for annual plants, like bluebur and small lupine, and warm, moist daytime hideaways for tiger salamanders and shiny, low-slung lizards called prairie skinks.

Plains pocket gopher

The "pocket" in the pocket gopher's name refers to the fur-lined pouches on either side of the face, in which the animal carries loads of cut-up stems and roots. The same characteristics are repeated in miniature by the pocket gopher's relatives, the pocket mice. These dainty creatures (tail and all, they'd fit into the palm of your hand) also excavate shallow tunnel systems and seal themselves in with dirt, but unlike pocket gophers, they venture out under cover of night to search for food. Their primary resource is seeds, which the prairie showers on them like the proverbial rain of gold. A stand of little bluestem, for example, will typically produce about 200 pounds of seed per acre (roughly 200 kilograms per hectare). To harvest this bounty, the sleek olive-backed pocket mouse (typically found on thin, sandy grasslands across the northern plains) sits up on its hind legs and sifts seeds out of the dirt, then stuffs them into its pouches with its forepaws. Weeds like Russian thistle and pigweed and native plants like butterfly weed and June grass are among its favorites. Seeds that aren't eaten immediately are stockpiled underground. In the fall, the animals move their caches below the frost line and hole up, drowsy but still conscious, for as long as six months. During all this time, they do not have to drink because they can extract water from their diet of dry seeds. Ord's kangaroo rat, a bigger, bouncier, desert-adapted species found in the Great Sandhills of Saskatchewan and other ultradry grasslands, shares this metabolic talent.

But of all the rangeland rodents—pocket mice, harvest mice, jumping mice, pygmy mice, ground squirrels, voles, and more—none is more remarkable than the black-tailed prairie dog of the western plains. Before the prairies were settled, this was the prairie dog's world. Its population ran into the billions, and if all the colonies, or towns, had somehow been amalgamated into one great metropolis, they would have occupied an area as big as Texas. In reality, prairie dog towns were, and in places still are, dotted across the short- and mixed-grass prairies—moonscapes of closely cropped grass pockmarked by craters of bare earth. Each of these earthworks marks the entrance to a shaft that plunges 6 to 10 feet (2 to 3 meters) below the surface and then levels out to lead to a complex of nest chambers and latrines. Prairie dogs descend into their burrows to sleep, give birth, nurse their young, and escape from predators, but they spend much of their lives on their earthworks, sprawled out on their bellies in the sun or upright and alert, ready to bark a warning at the first sign of danger.

On a warm summer day, a prairie dog colony is alive with the hustle and bustle of urban life, as youngsters chase each other, neighbors meet and "kiss," heads pop in and out of holes, and individuals waddle away from their mounds to graze on grass and forbs. But this apparently free-wheeling sociability masks a complex social order based on kinship. Each prairie dog lives in a small, tightknit family group, or coterie, that usually consists of one adult male, three or four breeding females, and nonbreeding youngsters. Young males disperse as juveniles to seek acceptance in another group, but females generally settle down in their natal coterie, close to their mother, sisters, and aunts. Although coterie members defend their space against intruders from neighboring clans, they live companionably among themselves, sharing burrows and other resources—with one significant lapse. During the birthing season, nursing females vigorously defend their nest burrows even against their kin, and for very good reason. The greatest danger to infant prairie dogs is their adult female relatives, who nip into unattended dens and kill the helpless huddles of pink pups. Yet once the pups emerge from their burrows, harmony is restored and adult females take turns nursing youngsters that, only a few days before, they would have attacked with vigor.

The reason for this turnaround remains obscure, but it may have something to do with a seasonal change in the "value" of the pups. In early spring when browse is scarce, another female's offspring may be most attractive as a source of nourishment. But by the time the pups appear above ground, their status changes abruptly from victim to defense. With her own progeny now exposed to predation, each female attempts to draw extra pups around her as protection for her own. If a predator attacks, there is a chance that her offspring will avoid harm by losing themselves in the crowd.

In the last 150 years or so, any threat posed to prairie dogs by their own infanticidal bent has been overshadowed by a greater and more persistent threat. To many cattlemen, a prairie dog colony, with its spreading expanse of close-clipped plants, represents little more than a terrible waste of grass, a view that has traditionally been seconded by range scientists. In 1924, for example, a team of respected researchers stated categorically that the prairie dog "does not possess a single beneficial food habit [beneficial to cattle, that is]; nor is there any argument . . . against its complete eradication on all grazing ranges." With that end in mind, prairie dogs have been poisoned with bait, suffocated with gas, drowned, and shot. Thousands more have been vacuumed out of their dens so that they could be relocated or, more often, offered a "humane" death. The result of this relentless and sometimes gleeful campaign of extermination has been a 98 percent reduction in the population of black-tailed prairie dogs since 1900.

In 2000, the U.S. Fish and Wildlife Service determined that the black-tailed prairie dog merited listing as a "threatened" species under the U.S. Endangered Species Act but then decided that "further action to place it on the list is precluded by actions to address higher-priority species." Dissatisfied with this ruling, conservation groups petitioned for a new assessment in 2004, in 2007, and again in 2008, without success. Fortunately, the plight of the black-tailed prairie dog has been recognized by conservation agencies in several Great Plains states and in Canada, where it is listed as Threatened. All the same, state agriculture departments in Montana, North Dakota, Wyoming, and Colorado continue to target prairie dogs as "pests," and one hunting-industry promotion recommends shooting prairie dogs as "a great way to get a child to be more comfortable with their firearms."

The war against prairie dogs has been based on the premise that they destroy rangelands. And it is certainly true that a prairie dog colony—with its patches of exposed ground and stumpy, chewed-off plants—bears a striking resemblance to a badly managed pasture. But recent research suggests that while prairie dogs can prevent the plants from recovering by keeping them snipped down, the rodents are rarely the primary cause of overgrazing. Because they cannot move into tall stands of dense vegetation, they often follow where heavy grazing by cattle (or formerly bison) has opened up a path. Ranchers who overtax their rangelands have effectively laid out a welcome mat and invited the prairie dogs to "infest" their land.

The claim that prairie dogs strip the prairie clean and leave nothing for larger herbivores to eat also turns out to be mistaken. In fact, far from avoiding prairie dog towns, bison are attracted to them and, given the chance, will spend a disproportionate amount of time grazing on the rodents' well-maintained lawns of tender young blue grama and buffalo grass. Although prairie dogs do eat plants that bison, or cattle, might otherwise consume, they compensate, at least in part, by maintaining patches of highly nutritious, highly digestible fodder. Presumably this fact explains why cattle that share the range with prairie dogs often gain weight nearly as fast as those that have the pasture to themselves. Although the arithmetic varies from case to case, the monetary rewards of removing prairie dogs is usually almost offset by the expense of "controlling" them.

Still, that "nearly" and that "almost" open up a rift, a classic conflict between economics and environment. Important new research being conducted on the Standing

Rock Sioux Reservation in South Dakota suggests that it is possible to meet the demands of both, by reserving the highest-quality pasture lands for grazing and permitting prairie dogs to thrive in sparser, less productive patches of prairie. This pragmatic strategy would reward ranchers with cash in hand and provide a zone of survival for a species of biological and cultural importance.

The ecological benefits of sustaining prairie dogs are beyond debate. Prairie dogs are a keystone species, the beavers of the grasslands. Simply by going about their lives, they shape and reshape their surroundings, thereby creating habitat for an astonishing variety of other creatures. Through their constant gnawing, they not only stimulate a fresh growth of green grass but also increase the diversity of the vegetation by opening up space for forbs such as scarlet mallow (or globemallow, as it is also called) and prickly pear cactus. These plants, in turn, provide food and shelter for other herbivores, including insects, birds, mice, rabbits, and hares. Some of these animals—cottontails, for example—are also attracted by the shelter of the prairie dogs' mounds, where they can tuck themselves out of the sun for hours at a time. Burrowing owls just move right on in, laying claim to abandoned burrows (usually on the outskirts of a prairie dog town) and building their nests underground. In the hazy heat of midsummer, the little owls can sometimes be seen standing, swivel-necked, at their burrow mouths or flapping languidly over the colony in search of grasshoppers.

More than 160 species of vertebrates have been observed on prairie dog colonies. Although some are drawn by the opportunities that prairie dogs create, others are attracted by the prairie dogs themselves. The plump little rodents figure in the diets of a host of predators, including two species of rattlesnakes and a dozen different kinds of hawks, plus weasels, skunks, badgers, bobcats, coyotes, and foxes. Prairie dogs also are, or at least once were, the sustaining resource of what is now one of the rarest carnivores on Earth. The black-footed ferret—a sinuous, mink-sized weasel with a black-bandit mask—is built to slink through prairie dog burrows and capture the rodents when they least expect it. Not surprisingly, as prairie dog populations dwindled, the ferrets also became increasingly rare. To make matters worse, the ferrets are also subject to canine distemper and to plague, an introduced disease that is endemic in prairie dog colonies. Under this malign confluence of influences, the species has come within a heartbeat of extinction. By the 1980s, the known population consisted of eighteen individuals in a single prairie dog town in Wyoming. Now, all these years later, the descendants of this remnant population are being bred in captivity and reintroduced to selected prairie dog habitat across the plains, from Saskatchewan to Mexico. Although their future remains precarious—the threat of disease is always there—the wild population has grown from zero to around three hundred.

Black-footed ferret

PRONGHORNS: OUTWIT,
OUTPLAY, OUTLAST

BACK IN 1872, a homesteader named Brewster Higley sat on the banks of Beaver Creek in Kansas and sang a tribute to his new home on the range, a land "where seldom is heard a discouraging word / And the sky is not cloudy all day." Since then, however, "discouraging words" about the future of the prairie's unique plants and animals have been heard all too frequently. In the last quarter century, for example, burrowing owls have essentially disappeared from the Canadian prairies, despite concerted efforts to sustain them, and declines have also been detected in many parts of the American plains. Similarly, the populations of mountain plovers (pale killdeerlike shorebirds that nest right out on the bald-headed prairie, often on prairie dog towns) have plunged by more than 60 percent in the last forty years. Much-loved birds, like prairie chickens, have been declining steadily, and signature species, like meadowlarks, have also been losing ground. "Less" and "fewer" have become our watchwords.

Happily, some species have bucked these dismal trends; the pronghorn is chief among them. Once as fantastically abundant as bison, with an estimated precontact population of 35 million head, the species was decimated by overhunting during the settlement era. By 1900, the herds had been reduced to a tattered remnant of some fifteen thousand. But a determined conservation effort turned the situation around, and the herds have rebounded several-hundred-fold in the last century. The count currently hovers around a million animals.

The pronghorn is a survivor. Although dozens of related species once roamed the Great Plains, only one, *Antilocapra americana,* made it through the mass extinctions at the end of the Ice Age. In so doing, the pronghorn permanently outdistanced its most fearsome predators, including North American lions and dire wolves; yet even today, it flies across the prairie as if they were still in hot pursuit. The fastest animal in North America, and possibly in the world, it can hit speeds of 60 miles (100 kilometers) an hour and maintain a blistering pace for several minutes. Running open mouthed, it gulps air down its extralarge windpipe and into its extralarge lungs, as its powerful heart keeps its muscles fueled with oxygen. This is a species with the prairie in its blood, the animal world's answer to wide open spaces.

Even the normally stolid Captain Clark—of Lewis and Clark fame—was moved to poetry by the pronghorns' grace and speed, which put him in mind of "the rapid flight of birds." Other observers were reminded of gazelles and jumped to the conclusion that pronghorns were a kind of antelope. (In fact, antelope are bovids, like bison and cattle; pronghorns are the only members of their own unique family and genus.)

Naturalists have found the pronghorn's social life even more deceiving. Although pronghorns of all ages and both sexes congregate during the winter, the herds break up in the spring, as the animals spread out across the prairie. Mature males—the big guys with the forked black horns—often lay claim to one or two sections of rangeland, which they defend against other bucks. Meanwhile, the does—the ones with the straight, short horns or occasionally none at all—gather in small herds that move unhindered from one male territory to another. Or so it goes until late summer when, with the approaching rut, the mature males begin to aggressively herd the does, as they stake sexual claim to their "harems." At least, this is what early observers thought was going on.

What actually happens turns out to be far more mischievous. Although the males do their best to keep tabs on the does, they are not always successful. If a female can't break away from the buck with a feint and a bound, she may outwit him by urinating in front of him. Distracted by her deposit of intoxicating scents, he indulges in flehmen, or intense sexual sniffing, while she makes a run for it. In the weeks just before estrus, a female may visit several breeding males before consenting to mate, apparently basing her decision on the bucks' relative ability to protect the herd from harassment and their success in male-to-male combat. Sometimes a female sneaks away from the herd just long enough to attract a consort of nonterritorial, or bachelor, males and returns to parade them past the harem buck. Standing safely out of harm's way, she then watches the snorting, pawing, rushing, and chasing that ensues, before presenting herself to mate with the victor. Far from being complete masters of the situation, male pronghorns are in much the same position as grouse on a lek as they strut their stuff before a discerning audience.

If the hidden quirks of pronghorn society are only beginning to be understood, one aspect of their biology has been clear from the outset. Although pronghorns are native to grasslands, they do not eat very much grass. As a result, they are perfect companions to grass-eating species such as bison and cows and do not interfere with the commercial production of cattle. In lieu of grass, pronghorns prefer a diet of shrubs and forbs, using their small, dainty muzzles to nip off the choicest morsels. This ability to forage selectively, one mouthful at a time, is important because many of the pronghorns' favorite foods are laced with poison. In the long evolutionary struggle between herb and herbivore, plants have fought back with every weapon at their disposal. Just as grasses contain grit that wears down herbivores' teeth, so many grassland forbs contain toxins that can cause indigestion, liver damage, miscarriage, and other bodily harm. (Locoweeds, lupines, and larkspurs are just three of a long list of prairie plants that regularly kill cattle.)

But through evolution, many prairie herbivores have responded to this threat by developing the ability to neutralize certain of these toxic agents. For example, sage and sagebrush contain turpentinelike compounds that can interfere with the microorganisms in an ungulate's complex stomach. Cattle avoid sages, and even mule deer can't eat too much, but pronghorns have completely mastered the challenge. Sage and sagebrush are among their staple foods and feature in their diet throughout the year.

WILE E. COYOTE AND CO.

THE FIRST RULE of ecology, on the rangelands as everywhere else, is that everything and everyone eventually gets eaten. But because organisms expend so much energy in daily living, only a fraction of what they ingest is converted into living tissue, or biomass. The proportion that is available for transfer to the next trophic level is usually no more than 10 percent. As a result, it takes a vast span of grass to support

STRANGE COMPANIONS

Coyotes and American badgers are middle-sized carnivores that share a taste for ground squirrels, prairie dogs, and other middle-sized prey. One might expect, then, that prey taken by one species would be a loss to the other. The fact that badger populations increase when coyotes are removed suggests that this kind of competition actually does occur and that badgers are typically the losers.

Under certain circumstances, however, the rules of engagement appear to shift from winner versus loser to win-win. If the two species find themselves working the same rodent colonies, and if such prey is scarce, the hunters may form an unlikely alliance. As the badger pursues its prey by burrowing underground, a coyote stands guard up top, ready to pounce on any rodent that makes a run for it. Meanwhile, down below, panic ensues, and the hard-working badger takes advantage of the confusion to make its kill.

Although hunting partners never share food with each other, they sometimes take time out to play—even to wrestle like pups—or simply to touch noses or lie down with each other. One observer described an encounter in which a coyote laid its head on the badger's neck, licked its face, jumped in the air, and made "other expressions of unmistakable joy." No wonder the two predators are often described in the stories of Indigenous peoples as neighbors and friends and are sometimes even said to address each other as cousins.

a herd of herbivores, which in turn can sustain only a handful of predators. Yet despite their relative scarcity, predators are present on the rangelands in every conceivable size and shape, from eight legged to six legged to no legged. Think of spiders, tiger beetles, ladybugs, assassin bugs, robberflies, and wasps; think snakes, lizards, turtles, frogs, toads, owls, and hawks; think shrews, weasels, skunks, and wild cats. Time was when the catalog would also have included wolverines, cougars, grizzlies, American black bears, and wolves, but those species were pushed off the plains during the settlement era. Wolves in particular were targeted for extermination with strychnine, traps, and guns, as an insult to commerce and a threat to civilization. (Today, although young wolves sometimes migrate across the prairies from the mountains and the north woods, they are inevitably shot or hit by a car before they have a chance to settle in.)

In the Bad Guy/Good Guy paradigm of the nineteenth century, the eradication of wolves should have been pure gain. Instead, it merely opened the way for a new breed of "outlaw" predator to move into the vacated space. As long as wolves were the dominant canids on the prairie, their slightly smaller cousins, the coyotes, didn't stand much of a chance. Any coyote that entered a wolf territory was likely to meet a bloody end, a victim of the cold logic of interspecific competition. As a result, coyotes were generally restricted to the gaps, or buffer zones, between packs of wolves and, though widely distributed across the plains, were nowhere plentiful. At least that is the impression one gets from the accounts of early explorers, who seldom mention seeing coyotes, or, to use their phrase, "bush wolves." By contrast, both red foxes and the diminutive (pussy-cat-sized) swift foxes

appear to have been widespread and common in some localities, perhaps because they were too small to compete with
wolves and thus were allowed to go their jaunty way without
interference.

After wolves were removed from the scene, the coyotes
muscled in, and the balance of power among canids began
to shift. Coyotes are versatile predators. Depending on the
circumstances, they can band together in packs to take deer
and compete with wolves, or they can hunt alone to catch
rodents, rabbits, and other small prey much as foxes do. So
just as wolves had once impinged on coyotes, coyotes now
aggressively took steps to limit the populations of their
smaller competitors. Red foxes, in particular, took a major
hit and, by 1900, had essentially vanished from large areas of
the grasslands. As the agricultural landscape changed, from
the 1930s on, red foxes pushed back in from the east and
reoccupied the region.

A superabundance of coyotes may also have been a pre-
cipitating factor in the decline of the swift fox. The smallest
of the wild dogs, these guileless little creatures were once
common across the short- and mixed-grass prairies, from
Canada to Texas and from the Rockies to the Dakotas. By
the 1940s, however, the species had been extirpated from
the northern plains, probably as a result of coyote predation,
together with loss of prey (notably rodents), loss of habitat to
the plow, and loss of life to poison baits intended for other
predators. Thanks to decades of effort by conservationists,
viable populations of swift foxes have been reestablished on
the Blackfeet Reservation in Montana, in southern Alberta
and Saskatchewan, and in South Dakota. Since the first
captive-bred animals were released in 1983, the population

in Canada alone has risen to around 650, and a thousand more have set up housekeeping in north-central and north-eastern Montana. These days, the greatest impediment to the survival of swift foxes appears not to be direct predation by coyotes (though, of course, that still occurs) so much as competition with their canny red cousin. When red foxes are present, swift foxes make themselves scarce, though exactly why this happens, no one can say for sure.

One thing we can be sure of, however, is that Wile E. Coyote still reigns supreme as top dog across the prairies. After a century or more on agriculture's Most Wanted list as live-stock predators, coyotes are still out there everywhere you look, howling, breeding, birthing, going about their business. By any sensible measure, attempts to limit their numbers have been a complete flop, much like trying to empty the bathtub without turning the water off. To have any lasting effect, a coyote-control program must target at least 75 per-cent of the breeding animals in a given neighborhood, a goal that is seldom achievable in practice. Yet without such drastic reductions, "control" becomes merely a means of thinning the population and ensuring that the survivors have adequate food to put them in prime condition to reproduce. When food is abundant, mature females produce large litters, and even subadults under the age of two are likely to breed and make a contribution to the baby boom.

But if coyote "control" can't be relied upon to achieve its intended results, it inevitably triggers a cascade of subtle, unseen, and unanticipated changes. For example, a reduction in the numbers of coyotes (however temporarily) provides an opening not only for red foxes but also for skunks, bob-cats, and other middle-sized predators, many of which prey heavily on duck nests. Fewer coyotes thus translates indirectly into fewer ducks and a chain of unknown effects on wetlands.

Meanwhile, on the grassy uplands, the removal of coyotes can also cause unexpected shifts in the abundance of rodents. For instance, one study in Texas found that when coyotes were present in good numbers, the prairie was home to flourishing populations of a dozen species of microherbivores, including deer mice, pocket mice, harvest mice, and Ord's kangaroo rat. But after half of the coyotes had been shot, the mouse populations crashed and the only species that remained was the kangaroo rat. (Presumably, predation by coyotes normally kept kangaroo rats in check and allowed the other, less competitive species to claim living space.) Since kangaroo rats are aggressive foragers that can push grasslands over the brink and encourage the invasion of desert shrubs, their dominance may have made the rangelands less hospitable for cattle. Meanwhile, with the coyotes held in abeyance, black-tailed jackrabbits multiplied to such an extent that they consumed any cattle fodder that was left. Thus, far from being dispensable animals (as one range manager described them in 1973), coyotes are a key species in rangeland ecology and the unlikely guardians of the cattle industry.

Long honored as the Trickster, Coyote still walks the prairie with us, calling us to look beyond the obvious.

CHAPTER 6

WATER OF LIFE

Here is a land where life is written in water.
THOMAS HORNSBY FERRIL,
INSCRIPTION IN THE COLORADO STATE
CAPITOL ROTUNDA, 1938

IT WAS LITTLE more than a puddle at the side of the road, the kind that appears in spring and is gone by midsummer. Two days earlier, it had lain frozen, burdened by a freak snowstorm. But in the sudden warmth of late April, the tide of the seasons had turned and the hollow was filled with sunlight and the bright creaking of frogs.

The far bank bristled with willows and wild roses and, poking up out of their midst, a single, stunted aspen that carried two large masses of twigs. One was flat topped, like a crow's nest, and did not appear to be in use, but the other was massive and messy and comically overbuilt for the meager tree that supported it. It had to be a hawk's nest. And, yes, there was a bird on this one, a large, dark form with a knob of a head that popped up above the rim. Do you suppose that it's a Swainson's hawk—the kind with the rich chestnut-brown bib—or its rarer cousin, the pale-fronted ferruginous hawk? Out come the field guides and binoculars. But wait: the shape that is sprawled on the hawk's nest doesn't look like a raptor at all. It's some kind of gray-backed bird with a sinuous black neck, a bar of white on its cheek and a stubby, ducklike bill that extends languidly over the rim of the nest. Could it be—what else could it be? A Canada goose!

I can't tell you if that goose had merely settled on the nest to sunbathe or if it had inexplicably taken a notion to lay and incubate its clutch at the top of a scrubby aspen. Strange as it may seem, this turns out to be a credible possibility. Although Canada geese usually nest on the ground near water or, at most, on slightly elevated surfaces such as grassy hummocks and beaver lodges, they sometimes do take over abandoned hawk nests as much as 6 yards, or meters, above ground. When the goslings are ready to leave, they step off the nest platform and flutter their tiny wings toward what will hopefully be a soft landing. But if I am not sure what that goose was doing, there is one thing I do know for sure. Even the most unprepossessing prairie puddle is a site of wonderment, a place where the unexpected happens daily and where nothing can safely be taken for granted.

It isn't just the life of watery places that is so interesting. Water itself is amazing. That raindrop on your nose was, until recently, a cloud. And before that, those same molecules of water may have gone through countless passages from snowfall to spring runoff, groundwater to marsh, marsh to river, river to ocean, ocean to wind, wind to cloud, in an endless, cyclic journey through the physical world.

THE TURTLE CONTINENT

There was a world before this world, Lakota storytellers
sometimes say, but the Creating Power was not pleased with
it. So he took his sacred pipe and his pipe bag, and he sang four
songs. With each song, the rain came, harder and harder, until
everything was flooded, and all the creatures were drowned.
Then the Creating Power unwrapped his pipe and opened his
pipe bag, which contained all manner of animals and birds. He
selected four animals known for their ability to stay under water
for a long time. One by one, he commanded them to dive and bring
up a lump of mud: the loon, then the otter, then the beaver. But they
all came up empty-handed. Only the turtle was left. It stayed down
so long the other animals were certain it must be dead. But when
it finally surfaced, its feet and claws—even the space between its
upper and lower shells—were filled with mud.

The Creating Power began to sing as he shaped that mud
with his hands. Eventually, he made so much land that the water
was pushed aside. "Water without earth is not good," thought
the Creating Power, "but land without water is not good either."
Filled with pity, he wept for the earth, and his tears became oceans,
streams, and lakes. "That's better," he thought. He gave the people
the pipe to live by, and told them all would be well if they lived in
harmony. He named this new land the Turtle Continent because
it was made from the mud that the turtle had brought. (This is an
abbreviated version of the story as it was told around 1910.)

And that is still not the whole picture. For not only is water capable of shifting from solid to liquid to gas—making it the only substance to exist in all three states under normal conditions on Earth—it also routinely makes the unfathomable leap from nonliving to living. The miracle of life derives from a thin, protoplasmic soup that, depending on the function of the cell, consists of between 50 and 90 percent water. (Our own tissues average out at about 50 percent.) So that same raindrop, which by now is dribbling off your chin, has likely also made a journey through the living world. From rain to soil, root to leaf, leaf to goose, goose droppings to soil, soil back to air, it has flowed through the food web, moving freely from organism to organism. Has it experienced life as an amoeba? An earthworm? A brontosaurus in a swamp? In the shape-shifting world of water, stranger things have happened.

Water is the medium of life, so it is not surprising that aquatic habitats—including oceans, rivers, lakes, ponds, and marshes—are among the richest and most productive environments on Earth. But it may be surprising to discover that this generalization holds true for the landlocked Great Plains, a region that we typically associate with drought, dust storms, and water shortage. On a planet where two-thirds of the surface is covered by water, the wide expanse of the prairies is a sea of dry land, broken here and there by small islands of aquatic habitat. Broadly speaking, the area covered by water is greatest in the eastern, tall-grass country and diminishes toward the west, following the logic of declining precipitation. Thus, in parts of central Manitoba, the inrush of moisture-bearing air from both west and south produces lush meadows and extensive, reedy marshes that occupy as much as 25 percent of the landscape. But on the prairies of Alberta, where the mountains block out the rain, wet meadows and marshes cover a mere 5 percent of the range. Similarly, if you cross the border into the States and head for the southern plains, the area of surface water declines incrementally. In Nebraska and Texas, for example, where the summer sun has the power to suck whole rivers out of their beds, the coverage of water declines to a fraction of a percent. Yet, though limited in physical area, the aquatic habitats of the prairies are biologically immense, capable under ideal conditions of producing more life in less space than almost any other environment on the planet.

PUTTING MARSHES ON THE MAP

FOR THE PURPOSES of discussion, it is useful to divide the aquatic habitats of the Great Plains into two categories: wetlands, or marshes, and rivers and lakes. Scientifically speaking, the term "wetland" refers to any shallow depression—less than 2 yards, or meters, deep—that is capable of holding water for long enough to permit the growth of sedges, cattails, and other nonwoody, semiaquatic vegetation. In other words, it is basically a puddle filled with water-loving plants. Yet this deceptively simple definition leaves room for almost infinite variability in water depth, temperature, permanence, chemistry, soil type, fertility, life history, and just plain chance, so every prairie wetland is, to some extent, a unique habitat that supports a unique community of organisms. *See Map 10: Wetland Regions of the Great Plains.*

Millions of these mostly small, self-contained marshes are dotted across the Great Plains, providing resource-rich refuges for a remarkable diversity and number of insects, reptiles, amphibians, mammals, and—most spectacularly— birds. For example, the Llano Estacado, or Staked Plains region, south of Amarillo, Texas, is so emphatically dry and flat that it cannot sustain the flow of a permanent river. Once described as the "great Zahara of North America," the region is now essentially a wall-to-wall expanse of cotton fields under irrigation. Yet this dreary prospect is redeemed by the presence of around twenty thousand shallow, saucer-shaped depressions known as playas, or playa lakes, that serve as havens for wildlife. (An additional roughly six thousand playas occur farther north in the Texas Panhandle, as well as in adjacent parts of New Mexico, Colorado, Kansas, and Oklahoma.) Scoured out by some still-mysterious interaction between wind, water, and underlying geology, these clay-lined basins have been in existence for at least ten thousand years and, judging from remains found buried in their beds, once served as watering holes for both Columbian mammoths and the Folsom-era humans who hunted them. Somewhat more recently, in 1541, when Francisco Vásquez de Coronado marched across these plains, he was as impressed by the playas—"round like plates, a stone's throw wide"—as he was by the bison that he encountered near them.

Throughout their long history, playas have presumably functioned in the same way as they do today: as catch basins for runoff and rain. As a rule, they fill up during spring cloud-bursts, reach peak levels in June, and thereafter (through evaporation and slow seepage) gradually disappear. The deeper the depression and the more water it holds, the longer the playa persists through the season. Since most playas are flat bottomed and less than a yard, or meter, in depth, they are generally dried out by December, at the latest. Only the deepest basins—including the increasingly large number that have been modified to retain water for irrigation—hold moisture year round and during periods of drought. Yet despite their ephemeral nature, playas sustain an abundance of life, including not only semiaquatic plants such as barnyard grass, curly dock, and pink smartweed but also a mix-and-match selection (differing from pond to pond) of up to six species of waterboatmen; five types of dragonflies; fifteen kinds of small marsh flies, or midges; nineteen kinds of freshwater shrimp; and dozens of other squirming, swimming, and flying invertebrates. This living soup, in turn, sustains populations of larger organisms, including amphibians (up to fourteen species of salamanders, frogs, and toads), reptiles (especially yellow mud turtles), and mammals (deer mice, cottontails, badgers, striped skunks, coyotes, and raccoons, among others).

But playas, above all, provide a haven for birds. During the breeding season, these unassuming wetlands serve as oases of nesting habitat both for marsh-adapted birds, such as ducks and red-winged blackbirds, and for a wide variety of typically terrestrial species, including short-eared owls, western meadowlarks, and ring-necked pheasants. The region also provides migratory stopover sites and wintering habitat for enormous numbers of waterfowl and shorebirds that breed elsewhere on the continent. Roughly two million ducks and geese crowd onto the playas each fall, including a gabbling contingent of mallards, northern pintails, American wigeons, and green-winged teal that nest in western Canada and the Dakotas. What's more, playas provide off-season accommodation to flight upon flight of gurgling sandhill cranes—more than 300,000 in all, or 90 percent of the midcontinent population—which land on the mudflats and in the dwindling ponds to feed on the thick slurry of crustaceans. And this is still to say nothing of the stilt-walking American avocets, with their upturned bills, or the Wilson's phalaropes, which spin round and round like tops to stir up food.

Little understood and vastly underappreciated, the playas of the southern High Plains are a vital ecological resource that helps to sustain waterfowl and shorebird populations over much of the continent. This function is shared by a number of other equally important wetlands in other parts of the Great Plains, many of them, like the playas, out in the middle of the prairies. The Central Kansas Wetlands, for example, include a variety of freshwater and saline marshes—notably Cheyenne Bottoms and the Quivira National Wildlife Refuge, both near Great Bend—that lead like stepping-stones through what is otherwise a vast plantation of corn, sorghum, and soy. Located on a major midcontinental migration route, these wetlands annually play host not only to waterfowl by the millions but also to half of all the shorebirds in North America. Although significant populations of shorebirds breed in Kansas and elsewhere on the Great Plains, a much larger number beat northward to the Arctic coast, only to breed, turn around, and head back south, often to wintering grounds on the other side of the equator. Hence the critical importance of refueling stops along the route.

Just north of the Central Kansas Wetlands lies the Rainwater Basin of south-central Nebraska, a rolling landscape dotted with several thousand playalike catchments. And then it's on to the Platte River, with its sandbars and wet meadows, and, a little farther along, the clear, sparkling—utterly incongruous—ponds that spangle the Sandhills of northwestern Nebraska. Filled to the brim by the Ogallala Aquifer (the vast reservoir of fossil water that underlies the south-central plains) and set against the silence of the grassy dunes, these wetlands are bursting with life and sound: the high-pitched complaints of long-billed curlews, the peeping of sandpipers, the metallic rattle of marsh wrens at nests in the cattails.

The final entry on the list of major wetland resources on the Great Plains is the Prairie Pothole Region, an area of some 270,000 square miles (700,000 square kilometers, about the size of Texas) that angles northwest from north-central Iowa through western Minnesota and the Dakotas and across the Canadian prairies from Manitoba to western Alberta. This entire swath of country is pocked with millions of mostly small, knee- to waist-deep depressions—called potholes, or sloughs—that are a legacy of the most recent glaciation. As the glaciers decayed, chunks of ice became buried at random under layers of gravel and silt, and when these blocks eventually melted, the ground collapsed, turning ice mounds into small meltwater basins. In keeping with this Ice Age heritage, prairie potholes still depend not so much on rainfall as on snowmelt, trickling over frozen ground, to replenish their supply of water.

Prairie potholes are proof of the adage that every little drop counts. Insignificant if considered in isolation, they gain significance when seen as elements of a sweeping, if fragmented, aquatic complex. The combined area occupied by potholes in the northern states, before settlement began, is estimated to have been 20 million acres (8 million hectares), giving them the same collective presence on the landscape as Lake Superior. And since 80 percent of the pothole region lies in Canada, this comparison barely hints at the wetlands' original extent, much less at their ecological significance. Vitally important as a source of protein for both resident and migrant birds, these weedy, reedy marshes are perhaps most famous as the "duck factory" of North America. Although they only amount to about 10 percent of the waterfowl-breeding habitat in North America, they typically produce between 50 and 90 percent of all the ducklings that fledge in any given year. Clearly, prairie potholes are spectacularly productive. The obvious question is, how do they do it?

HOW PRAIRIE POTHOLES WORK

AT FIRST GLANCE, a slough does not appear to have the makings of a biological powerhouse. What is it, in the final analysis, but a pool of stagnant water? Sunk into poorly sorted glacial rubble that inhibits organized flow, a typical pothole has only weak and fitful connections with the surrounding groundwater. Depending on the circumstances, any given pothole may receive water through seepage, lose it through underground leaks, or experience both input and outflow through different parts of the basin simultaneously. A slough that receives an influx of groundwater is also typically hit with a burden of dissolved salts picked up from the underlying geological formations. If this briny solution is permitted to escape from the basin either through overbank flooding or underground leaks, the salts flow out with the water and tend not to accumulate. But if the basin has no outlet, the water gradually evaporates, the concentration of salts builds up, and the wetland becomes saline (with chlorides) or alkaline (with sulfates). Some prairie potholes are ten times as salty as the sea, surely pushing the limits of habitability.

Sloughs also present a variety of other potentially life-threatening challenges. In the swampy warmth of mid-summer, as submersed plants die back and rot, the water becomes starved for oxygen—if there's still water left to starve. Like playas, prairie potholes are notoriously unstable, typically filling in spring and dwindling over a period of weeks or months, with little chance of renewal before the next spring thaw. How long water persists varies from pothole to pothole and from year to year, in response to such variables as the depth of the basin, flow to and from the groundwater, and the uncertainties of the weather. Depending on its duration, a slough may be classified as ephemeral, that is, a thin layer, or sheet, of water that lasts for a matter of days; temporary, a puddle that persists into early summer; seasonal, a pond that dries out annually but that often remains wet till fall; or permanent, a wetland that withstands all but the most severe droughts. (This last category includes such birding hot spots as Delta and Oak Hammock marshes in Manitoba, Last Mountain Lake and the Quill Lakes in Saskatchewan, and Beaverhill Lake in Alberta, all of which have been honored as Wetlands of International Importance under the UN's Ramsar Convention on Wetlands.) And these distinctions are all overridden by the insult of the northern winter, which closes in on wetlands regardless of their status and deprives them of light, warmth, and other necessities.

Yet the wonder is that, in one way or another, life goes on. For example, lack of dissolved oxygen is a crisis for many species that breathe through their skin (like flatworms) or with gills (like dragonfly larvae, or nymphs) and that accordingly are restricted to only the most permanent and well-aerated sloughs. But certain other species are adapted to cope with anaerobic or near-anaerobic conditions. Some daphnia, or water fleas, for instance, respond to oxygen-starved water by pumping up their production of oxygen-grabbing hemoglobin, an adaptation that not only

turns them a surprising shade of pink but also enables them to take full advantage of every available molecule of dissolved oxygen. Mosquito larvae, or wrigglers, by contrast, live in the water but obtain their oxygen by means of breathing tubes that poke into the air, on the same principle as snorkelers. Similarly, a variety of aquatic bugs, such as waterboatmen and backswimmers, rise to the surface and pick up

Backswimmer

loads of air, which they carry down with them like scuba divers. Often you can see them scooting around beneath the surface of the slough, all shiny and silvered with bubbles.

So far so good. But what happens when the very medium of pond life—the water— disappears? This is a routine crisis for prairie potholes (and other transient wetlands on the Great Plains) and one that, to a remarkable extent, evolution has turned to advantage. Through thousands of reiterations, the so-called wet-dry cycle has given rise to an ecosystem that not only is adapted to cope with severe hardship but depends on it. Here's how it works. If a basin remains full of water for even two or

Waterboatman

three years in a row, the community of life within it begins to deteriorate, becoming both dramatically less diverse and less productive. The first sign of this degeneration is a gradual die-back of cattails, bulrushes, and other emergent growth— those plants that are rooted in the muddy bottom, push up through the pond, and emerge crisp and spiky above the surface. With continuous inundation, this once-vigorous foliage either disappears from the basin over a period of months or else recedes toward the margins.

Although the causes of this die-off are often mysterious (aging rootstocks or lack of oxygen in the mud?), the process is sometimes accelerated by an assault force of muskrats. Where conditions are suitable—namely, in well-vegetated potholes that retain free water throughout the year—these plump little herbivores often increase to such extraordinary numbers that they "eat out" large areas of marsh vegetation. (They are basically aquatic field mice and just as fecund.) Their habitat destroyed, the rodents also crash, leaving the wetland doubly diminished.

One loss leads to the next. Fewer emergent plants means less habitat for all the little creatures that live on and around their leaves and stems, including algae, tiny crustaceans, snails, and insect larvae. In addition, the elimination of emergent vegetation allows the wind to stir through the pond, thereby muddying the water and reducing the amount of light available to plants that grow beneath the surface, including the slimy, undulating stems of sago pondweed and coontail. As this submersed vegetation begins to die off in its turn, the marsh completes its degeneration phase and reaches its lowest ebb—the lake phase—a stage that is characterized by large areas of open water and relatively low production of plants, microorganisms, insects, and everything that eats them. And what doesn't?

The slough persists in this depressed condition until it is revived by drought. To regenerate, it first has to dry out or, as the biologists put it, experience drawdown. Although the plants that are at home in prairie potholes typically require standing water to grow, their seeds are only able to germinate on dry land. So while the brittle mud of a dried-out slough may look dead, it is actually a green-and-growing marsh just waiting to happen. A small square of exposed slough bottom (a square yard, or meter, on a side, dug to the depth of your thumb) can be expected to contain somewhere between 20,000 and 45,000 seeds for as many as two dozen species of plants. The exact nature of this seedbank depends on the history of the pond and the species that set seed in it during the preceding wet-dry cycle. Some of them—like the barbed fruits of nodding beggar-ticks—have been lying there since the last time the slough dried out, anywhere from one to thirty years earlier. As a mudflat specialist and an annual, the nodding beggar-tick germinates at drawdown, produces a profusion of yellow blooms, and sets and deposits a stock of fresh seed. When water returns to the pond, the mature plants are drowned out, but the seeds of the future lie waiting in the ground.

For cattails and other emergent plants, by contrast, this same inrush of returning water is the stimulus for vigorous growth. Seedlings that germinated on the dry mudflat now sink their roots into the gooey muck, pushing out runners in every direction and sending up juicy young shoots. Then, once the marsh is fully reflooded, pondweed and other submergents begin to germinate in their turn, as the marsh flows into its regeneration phase. Meanwhile, various forms of animal life are also emerging from the mud, in a kind of ritual reenactment of creation. The first to crawl out of the ooze are water fleas and other tiny crustaceans that hatch from eggs, laid the last time the pond was wet, that have remained buried and dormant in the sediments. Some species are so deeply attuned to the wet-dry cycle that they cannot reproduce unless their eggs are first desiccated and then reflooded. This is also true of the "floodwater mosquitoes" (genus *Aedes*), which lay their eggs on plant stalks in dried-out meadows, leave them high and dry all winter long, and rely on spring flooding to revive them. The transition from dormant egg to flying, blood-sucking adult can be achieved in a mere ninety-six hours. Instant mosquitoes: Just add water.

Once the water begins to flood into the pond, life returns in a rush, as these pioneering species soak up the nutrients deposited during the preceding degeneration phase, or die-off. How long this surge of life persists depends, more than anything else, on how long water remains in the basin. If the slough only stays wet for a matter of weeks, the recovery is cut short and is limited to species that can hatch, mature, and deposit new eggs on a tight schedule. But if the wetland is seasonal or permanent—if it lasts for several months—the recovery is enriched by a wave of colonists. These are mobile species of insects, reptiles, amphibians, mammals, and birds that escaped from the drawdown by moving to deeper, more permanent sloughs and that begin to find their way back as the habitat improves. For example, many kinds of water bugs and beetles cope with the dying-and-reviving nature of wetlands by making regular reconnaissance flights, scanning the landscape for reflective surfaces of an appropriate tint and size. By responding to the glint of water, they can home in on regenerating habitat or, if their instincts happen to fail them, occasionally on the roofs of blue Hondas! Sometimes they bring stowaways with them, in the form of aquatic mites and crustaceans that have tucked themselves under a wing and that join in the reconstitution of the marsh community. Ducks and other waterfowl also provide transportation from pond to pond, carrying in algae, snails, and other clandestine cargo. (Of the two dozen species of snails found in prairie potholes, several can survive drawdown by burrowing into the bottom or sealing themselves inside a mucilaginous coating. Most, however, can't cope with dry conditions and have to be reimported with each successive cycle.)

With persistent inundation, the pioneering species—the ones that require drawdown to reproduce—gradually fade from the scene and are replaced by organisms adapted to long-term inundation. This category includes, for example, the larvae of tiger salamanders, large, slow-growing, tadpolelike creatures with a collar of fringed gills, which typically spend up to two years in the water before they metamorphose into gill-less, terrestrial adults. Indeed, some individuals avoid making this transition altogether and spend their entire lives in the pond, eventually reaching sexual maturity and even breeding in the form of oversized larvae, called paedomorphs. (Tiger salamanders frequently live to an age of eight to ten years on land, though little is known about the life span of the aquatic forms.)

Yet the high times can't last forever. Inevitably, whether in a matter of months or years, the conditions of life in the pothole will alter and the marsh will cycle on, passing from full flood to degeneration and from drawdown to rebirth.

INSIDE THE DUCK FACTORY

THE WET-DRY CYCLE provides a key to understanding how the Prairie Pothole Region is able to produce tens of millions of ducklings in an average growing season. The surge of life that follows drawdown, and the cycling of resources through the marsh, produces a periodic, self-renewing, superabundant crop of algae, minicrustaceans, insect larvae, and other high-protein food that is critical to the success of breeding waterfowl. Even species such as mallards and northern pintails, which are seed eaters for most of the year, switch to a diet of aquatic invertebrates during the nesting season. It is no small thing, after all, for a female duck to produce a clutch of nine or ten eggs, one a day for as many days, and then to start all over again if the first clutch fails to hatch. (Nest failures are common, and female mallards have been known to make as many as four nesting attempts in a single season.) Because regeneration is usually triggered by snowmelt in the spring, the pulse in the production of pond life is ideally timed to support egg laying and to sustain the hen through the month-long vigil of incubation. And there is still plenty to eat in midseason, when the ducklings peck their way out of their shells, follow their mother to a slough, and bob along after her in search of food.

What's good for ducks is also good for the people who admire them. The delight of traveling through pothole country, especially in a wet spring, derives not only from the numbers of ducks but from their variety. If one pond sports a pair of mallards, the next one will surprise us with blue-winged teals or American wigeons or lesser scaups, each appropriately dapper or dappled or iridescent. Here's a pair of northern shovelers, slowly swaying their big, trowel-like beaks through the shallows, with much the same motion as if they were wielding metal detectors. (In fact, the birds are feeding on microscopic pond life, which they obtain by passing water over their tongues and out the sides of their bills, which are equipped with comblike filters.) In the next pond down the road, a pair of common mergansers may be swimming side by side, he resplendent in green, she with a tousled headdress of cinnamon, their beaks unexpectedly pointed and startlingly red. Hooked at the ends and serrated along both sides, these weapons permit mergansers to catch amphibians and (where the water is sufficiently oxygenated) minnows and other small fishes. No two potholes are likely to yield an identical list of species.

Northern shovelers

Common mergansers

The preference of particular species of ducks for particular potholes is a sign of the subtle-but-significant differences between basins. Through the intricacies of the wet-dry cycle, slight variations in depth and water duration are translated into tangible differences in the amount and types of invertebrates, or duck food, in each wetland. When this slough-to-slough variability is multiplied by millions of basins and extended across the northern plains, it results in a varied landscape of wetland habitats that, collectively, are suited to meet the needs of a varied array of duck species. Pair by pair, the birds select the potholes that are best able to satisfy their specific requirements. Alternatively, if they can't find one wetland that will do the entire job, the birds move from slough to slough, perhaps using one basin for feeding in early spring, another for nesting, and still another for rearing ducklings. This ability to provide a variety of quality habitat is another reason that the Prairie Pothole Region is such a hotbed of duck production.

Not surprisingly, ducks are more searching in their assessment of wetlands than human observers are, and their reasons for preferring one slough over another are not always apparent to us. Nonetheless, some general patterns are fairly easy to discern. Most ducks that nest on the prairies fall into one of two main groups, or tribes, each of which is adapted to a particular lifestyle. On the one hand are the dabblers or, as they are more formally known, the members of the genus *Anas,* a lineage that includes the mallards, teals, wigeons, gadwalls, pintails, and shovelers. Often seen upended, tail waggling, in the slough, these ducks are adapted either to feed near the surface of the water (like shovelers) or, by tipping up, to forage on the slough bottom. Accordingly, they prefer shallow potholes, usually with an interspersion of open water and emergent plants. Since their preferred habitat is highly dynamic—flooded one year, dry the next—they tend to be opportunistic in their choice of breeding grounds, with each pair settling in the first unclaimed slough that they happen to come upon during the spring migration. Finally, because their chosen potholes are not deep enough to offer protection from terrestrial predators, dabbling ducks often nest in dense vegetation well away from wetlands and lead their ducklings on a long trek back to water.

The other tribe of pothole-nesting ducks is the divers. Less abundant and perhaps less familiar than the dabblers, this group includes two eye-teasing pairs of lookalikes—the canvasback and redhead, and the ring-necked duck and lesser scaup—all of them members of the genus *Aythya*. An unusual assortment of other, less closely related species can also conveniently be mentioned in the same breath, notably, common goldeneyes, buffleheads, mergansers, and ruddy ducks. (Ruddy ducks, in particular, are worth checking out, if only for the bright blue bills of the breeding males.) What these birds all have in common is their ability to dive, swim underwater, and exploit the resources of the deeper and more permanent prairie marshes. Unlike dabblers, which prefer ephemeral and seasonal basins, divers tend to be found in seasonal and permanent potholes with areas of open water. Although several of the diving species protect their eggs by nesting in hollow trees, others find safety for their broods on islands or in stands of emergent plants in the middle of the basin. And since their chosen waters are relatively reliable, as prairie potholes go, diving ducks tend to return to the same nesting sites year after year, rather than playing catch-as-catch-can in the way that dabbling ducks do.

Reliable, as prairie potholes go. The fact is, of course, that even the deepest and most permanent wetlands cannot be depended upon. A succession of snow-free winters and hot summers can dry the whole region out, including even large and persistent marshes. Clearly, both dabbling and diving species are adapted to thrive in these dynamic environments when conditions are "ducky," but what happens when they're not? How do waterfowl cope when their habitat dries up?

The short answer is, they don't. Instead, like other types of pond life, ducks have little choice except to get by as best they can. Most species of pothole-nesting ducks tend to overfly the dried-out potholes and continue on to the north, often settling down in the boreal forest. In this less-than-ideal environment, they produce a meager crop of young, often too few to sustain the breeding population. Thus, fewer sloughs translates directly into fewer ducks. This relationship between the number of ponds in the pothole region in May and the number of breeding ducks tallied the following spring has been documented repeatedly in data that go back to the early 1960s in Canada and the mid-1970s in the United States. Although the graphs do not line up precisely, because of the many complicating variables that influence the ducks' success (including the timing of spring breakup, unseasonable blizzards, and so on), the parallel run of the lines from peak to trough is too striking to be questioned.

WHERE HAVE ALL
THE DUCKLINGS GONE?

THE BLEAKEST YEARS so far on record for North American ducks extended from the mid-1980s to the early 1990s, a period during which the climate wavered queasily between wet and dry for several years before settling into one of the most intense droughts since the Dirty Thirties. The dabbling ducks, in particular, were hit hard by this turn of events, as durable species like the mallard plunged toward the lowest numbers ever recorded. (In 1985, this most populous of North American ducks bottomed out at about five million birds, a worrying 40 percent below its long-term average numbers.) Worse yet, even during the occasional seasons when the prairie potholes did fill up, the ducks inexplicably failed to rebound. It seemed that some unknown difficulty was compounding the effects of the drought and preventing the ducks from recovering during the wet phase of the cycle. Alarmed that this downturn might be permanent, researchers embarked on an urgent search for answers.

One fact was immediately obvious. The Prairie Pothole Region was no longer the paradise for ducks that it had been at the beginning of the settlement era. Since 1900, somewhere between 20 and 85 percent of the original prairie potholes have been ditched, drained, dredged, filled, leveled, plowed, or paved; the exact percentage varies from district to district. (In general, losses have been lowest where land is used for grazing and highest on croplands and in urban settings.) Although the rate of wetland conversion has declined in recent decades—and the pace of restoration has picked up—the net balance is still negative. Between the late 1970s and the early 1980s, for example, an additional 1,500 square miles (4,000 square kilometers) of wetlands were appropriated for farming operations in the United States alone, and the losses in Canada were similar. This was the very period when the continental population of ducks began to slide into the basement.

Loss of wetlands is never good news for ducks, yet biologists were not convinced that it was a primary cause of the 1980s crisis. After all, the immediate problem wasn't really a lack of potholes; it was the inability of the ducks to ride the wave of recovery following drawdown and fully reoccupy the wetlands still available to them. Some link in the chain between the wet-dry cycle and duck reproduction had been broken, so an abundance of duck food during wet years no longer resulted in an abundant output of ducklings. Could it be, the researchers began to wonder, that something was suppressing duck reproduction? Studies conducted in the early-to-mid-1980s quickly confirmed this disturbing hypothesis. Instead of the 15 to 20 percent "hatchability" (one or more hatchling per nest) that is considered a bare minimum to keep duck populations at a steady state, the results from the Prairie Pothole Region were dismal: an average of 7 percent nest success for pintails, 12 percent for shovelers, 14 percent for gadwalls, and 8 and 11 percent for mallards in two different areas. These figures meant that even when the ponds were crowded with ducks and the nests were brimful of eggs, the vast majority of nesting attempts were failures. The famed duck factory of North America appeared to have gone bust, and a once-productive population was now merely going through the motions of reproduction.

One curious detail—a needle in the haystack of data—emerged from this research. The studies showed a strong and surprising relationship between the prospects of a given nest and the human land use in the countryside that surrounded it. Specifically, the greater the percentage of cropland in the area, the lower the likelihood that the ducks would be successful. This was partly because spring-seeding activities took a toll on nests that were situated in stubble fields, sites that are especially favored by pintails. But this kind of collateral damage did not appear to be a significant problem for other

ODD DUCKS

The northern pintail is a queer duck. A member of the dabbling duck tribe, it looks a bit like a mallard that has been stretched at both ends to produce an elongated neck and, in the male, an extravagant, quill-like tail. The drakes are also distinguished by a sleek chocolate-brown head, a pure white breast, and a swirled boundary where the colors meet on the neck. (If an Art Deco designer had set out to create a duck, it might well have resembled a male pintail.) Hens are clad in camouflage plumage the color of dry grass, plain but elegant.

What makes northern pintails unusual is their lackadaisical approach to the selection of nesting sites. Unlike most other dabblers, which conceal their nests in thickets of grass or shrubs, pintails settle down any old where, even in sites that are completely open. Under natural conditions, when their nests were mere points in unbroken vistas of grass, this strategy presumably made sense. A predator had to be lucky to find them. But these days, when the grasslands are broken into fields and fragments and liberally studded with trees, hawks and other predators have gained a troubling advantage.

Pintails that nest in crop stubble are at special risk, with only a 4 percent chance of success. (Nests are destroyed during spring seeding or hit by predators.) As a result, pintails are the only species of predominantly pothole-nesting ducks that have not rebounded from the 1980s drought. Hopefully, a shift in agricultural practices toward fall-seeded, or winter, wheat will give pintails a better chance by reducing spring disturbance and providing early-season cover for nests.

species of dabbling ducks, which nest in tractor-no-go zones such as ditches and patches of brush. The significant problem appeared to be not farming as such but one of its unintended consequences. In one way and another, agriculture had inadvertently boosted the populations of middle-sized predators, such as foxes, raccoons, and skunks. By the mid-1980s, this cadre of nest-raiding mammals (aided by bullsnakes in some localities) was making a meal out of both duck eggs and sitting ducks, and there didn't seem to be much that anyone could do to stop them.

Then two apparently unrelated events occurred. In 1985, as drought took hold of the prairies in earnest, the U.S. government introduced a program designed to prevent soil erosion and support the region's beleaguered farmers. Known as the Conservation Reserve Program, or CRP, this initiative provided an annual payment, on a ten- or fifteen-year term, to landowners who agreed to retire marginally productive land and plant it to grass, alfalfa, or other forms of soil-binding permanent cover. Conceived as a temporary response to hard times, the CRP has since been revised and renewed to buy protection for thousands of fragments of upland habitat in the Dakotas and adjacent parts of the Pothole Region. In so doing, the program almost inadvertently created a vast new resource of quality upland habitat for wild animals and birds. With better cover for their nests, ducks on CRP lands frequently enjoyed success rates in the range of 20 to 30 percent, more than enough to lift the populations out of their slump.

Meanwhile, at the same time as new habitat was being created, ducks received another boost when a decade of drought finally drew to a close. When the long-awaited rains returned in the early 1990s and the sloughs filled up, the

Prairie Pothole Region moved from drawdown to regeneration in a headlong rush. After a period when it had seemed that ducks were on their way down, they were suddenly present in every ditch and slough, and the populations of several species surged toward record numbers. For the first time in history, however, the pulse that replenished the population came mainly from the CRP lands in the Dakotas, rather than from the Canadian prairies, which had been the powerhouse of duck production in previous surveys. Thankfully, the recovery has been durable, with populations of most waterfowl species stable or on the rise in the early decades of the twenty-first century. The ability of pothole-nesting ducks to rebound from near disaster stands as a testament not only to the regenerative powers of prairie wetlands but also to the ongoing efforts of conservation-minded people.

PRAIRIE RIVERS

PRAIRIE POTHOLES AND other ephemeral wetlands are classic examples of ecosystems adapted to cope with almost constant disturbance. Faced with a relentlessly dynamic environment, pond life has become almost as fluid and responsive as water itself. And this same description also applies, though in strikingly different ways, to the life of the two major river systems that cut across the Great Plains. They are, to the north, the Saskatchewan and, to the south, the Missouri. *See Map 11: River Systems of the Great Plains.*

The Saskatchewan system extends across the Canadian plains like a spindly arm, with its fingertips in the ice fields of the Alberta Rockies and its elbow at the north end of Lake

Winnipeg. Rising as a thousand milky meltwater trickles that tumble down the continental divide, the water converges into creeks and then rivers that flow to the north and west and ultimately join to form the North Saskatchewan River. Meanwhile, water cascades down the mountains to the south and west into the valleys of the Oldman, Bow, and Red Deer rivers; together, they become the lovely, often lazy South Saskatchewan River. Cutting cross-country through the farmlands of Alberta and Saskatchewan, the two branches meet east of Prince Albert to form a single, gray-green flow, the Saskatchewan River. From there, it is on to Lake Winnipeg and then north, via the Nelson River, into the Arctic Ocean at Hudson Bay. From headwaters to mouth, the major rivers of this system cover a combined distance of some 1,600 miles (2,600 kilometers) and provide drainage to more than 400,000 square miles (a million square kilometers) of land in the three Prairie provinces.

The Missouri River system is, if anything, even more magnificent. From sources in the Bitterroot Range of northwestern Wyoming and southwestern Montana, the mainstream of the Missouri meanders east across Montana and then south through the Dakotas before emptying its muddy flow into the Mississippi River at St. Louis, for a total run of almost 2,500 miles (4,000 kilometers). Routed along the shore of a long-vanished Ice Age glacier, the great river snakes diagonally across the center of the continent, attracting drainage from both the north (through tributaries such as the Milk and the Frenchman) and the west (through, among others, the Yellowstone, Little Missouri, Niobrara, and Platte rivers). All told, the rivers of the Missouri system drain an area of about 530,000 square miles (1.4 million square kilometers), including the southernmost Canadian prairies and one-sixth of the continental United States. That's the entire sweep of country from the Rockies to northeastern Missouri and from southern Saskatchewan to central Kansas.

From central Kansas south, by contrast, the land slopes ever so gradually away from the Missouri basin and entrains the rivers to run directly into the Mississippi River or the Gulf of Mexico. These include the Cimarron, Canadian, and Arkansas rivers of the Arkansas River system, and the Brazos River of central Texas. Thus, in one way or another, the waters of the dry-and-dusty American heartland all end up in the Caribbean Sea and, thence, in the North Atlantic.

OF FISH AND FLOOD

ALTHOUGH WE TEND to think of rivers as if they were all alike—just an onward rush of liquid confined between banks—every watercourse is unique, distinguished by a particular set of physical characteristics. Does the water freeze solid in winter or flow freely throughout the year? Is it fast and deep or soupy and shallow? Clear as crystal or burdened with eroded soil? Does it shrivel away to nothing in midsummer? The answers to these questions define the conditions of life in a particular stream and determine the range of species adapted to survive in it. Thus, rivers like the North and South Saskatchewan and Upper Missouri, which are ice-bound for part of the year, support far fewer types of fish (some two or three dozen at most) than rivers farther south, like the Arkansas and lower Missouri (each with a species list of a hundred-plus). Similarly, the seasonally low and often oxygen-deficient waters of the arid western plains tend to have fewer types of fish than the more generous and stable rivers of the tall-grass prairies. In the eastward-flowing Arkansas River, for example, the species count more than doubles from west to east, rising from 54 to 117.

A river is not really one place but many, a long, sinuous linkage of different environments connected by the flowing fabric of the water. Even adjacent stretches of the same river can be remarkably dissimilar, from sparkling, well-oxygenated riffles where the water slips over gravel beds to slow, swirling pools behind logjams to marshy backwaters set away from the main channel. Here fish spawn or feed or seek refuge from predators. There frogs lay their eggs, turtles bask in the sun, and great blue herons stalk through the shallows. The diversity of life in the river depends on its ability to maintain a constant supply, or flow, of different habitats, so that a diversity of new opportunities is continually created.

For the major rivers in both the Saskatchewan and Missouri systems, the principal force of renewal has, until quite recently at least, been the chaotic ritual of spring flooding. As so-called large floodplain rivers, the North and South Saskatchewan and the Missouri are naturally characterized by highly dynamic flows and were never meant to be confined within narrow channels. In the case of the Missouri, to cite the best-documented example, the river followed a seasonal rhythm, falling to its lowest levels in winter, experiencing a small surge at breakup in spring, and then, depending on the year, often really kicking over the traces in early June. Forced by the onrush of snowmelt from the mountains, the river swirled up over its banks and out across its wide,

COOL AS A CLAM

Just when you thought you had heard everything, someone tells you about the unionid mussels. They're aquatic mollusks, or shellfish—often called "freshwater clams"—that live in creeks and rivers across the Great Plains and that produce the pearl-lined, ridged-backed shells sometimes to be found around muskrat lodges or strewn on beaches.

At first glance, the life of a mussel sounds like boredom in the extreme. The animal lies on the river bottom, half-buried in gravel or sand, and pumps water in one end of its body and out the other. But things start to get interesting when the mussels breed. The male does his part by releasing clouds of sperm into the water. When the female sucks in this seminal fluid, her fertilized eggs develop into larvae that attach to her gills. To mature, however, the larvae have to abandon mother and hitch a ride on a fish.

How to make contact? In some species, the female simply turns the larvae loose and lets nature take its course. But many mussels reduce the risk of failure by luring a fish in close. The female plain pocketbook mussel, for example, has a lip of tissue, or mantle flap, that looks astonishingly like a small shiner fish, complete with eyes and stripes. When a bass strikes at this wriggling lure, it ends up with a mouthful of mussel larvae, which hook onto its gills or fins and grow to maturity. Once fully developed (usually without harm to the fish), the young mussels drop to the river bottom, where if conditions are right, they settle into the humdrum routine of adult life.

flat-bottomed overflow valley, or floodplain. With the full force of a natural disaster, the water scoured away sandbars, uprooted shrubbery and full-grown trees, tore out logjams, and generally ran amuck, leaving the entire floodplain in chaos.

But all that mess and disorder held the seeds of rebirth. When the flood subsided, it revealed a soggy landscape of ponds and puddles, many of them linked to the river's main flow, that would quickly grow up into a lush and varied mosaic of oxbow lakes, marshes, and wet meadows. Amphibians laid their eggs in these quiet waters; certain fish, like the northern pike, splashed through the shallows to spawn; others relied on the floodplains as rearing grounds for their young. And all this productivity was accelerated by an input of nutrients from the rotting muck that was strewn around the countryside in the wake of the flood. In this way, a landscape of aging floodplain forests and silted-up wetlands was periodically restored to the freshness and vigor of youth, and the resources of the system were freed up to support a new burst of growth.

Meanwhile, as the main flow of the river fell back into its channel (around many a new twist and bend), its waters were muddied with silt and decaying vegetation. Yet this apparent burden constituted a food-rich resource for the community of plankton in the river and hence for plankton eaters, such as insect larvae, crustaceans, and fish, and ultimately for most other organisms that lived in the stream. This surge of productivity is known as the flood-pulse, and, though still poorly understood, it appears to be as important to the life of rivers as the wet-dry cycle is to marshes. For example, many species of river fish—including native prairie minnows like the

Arkansas River shiner and speckled chub—rely on spring flooding as a signal to spawn. By responding to this seasonal cue, the fish are able to coordinate their life cycle with the turning of the year, thereby ensuring that their young receive the full benefit of both the flood-pulse of productivity and the growth-enhancing warmth of midsummer.

In the last half century, however, this natural rhythm has been disturbed. In some instances, the seasonal flood cycle has been dampened by an overall lowering of the rivers' flow caused by heavy withdrawals for irrigation. (These demands have intensified since World War II, when pumps powered by surplus airplane engines were first hooked up to central-pivot irrigation systems.) As water is pumped out of underground formations, the level of the groundwater falls, just as if it were being sucked out by thousands of high-tech drinking straws. As the water table drops, there are no reserves left near the surface to trickle into the river, through seeps and springs, and help to keep its flow replenished. As a result, the river is now so diminished that it can no longer overflow, flood the surrounding plain, and replenish the groundwater through a trickle-down effect. Caught in this self-defeating cycle, the water table underlying much of the southern plains has dropped by anywhere from 5 to more than 150 feet (about 2 to more than 45 meters) since 1950, and the volume of many rivers is now drastically reduced. To cite one poignant example, the Middle Arkansas River in Kansas has dwindled from an average flow, in the early 1940s, of up to 1,500 cubic feet (40 cubic meters) per second to a maximum rate by 2010 of no more than 250 cubic feet (7 cubic meters) per second, a fall of about 85 percent. The hotter, drier climate that is coming our way is not going to help.

Not surprisingly, this degradation is taking a heavy toll on species that are adapted to live in free-flowing, seasonally flooding rivers. A century or so ago, that description fit most rivers on the southern plains, including the little Smoky Hill River of west-central Kansas. When the fish of the Smoky Hill basin were first cataloged in 1885, researchers hauled in specimens by the dozens. The list of species was evocative: hornyhead chubs, johnnydarters, brassy and suckermouth minnows, black bullheads, stonecats, green sunfish, blacknose shiners, southern redbelly dace, and many others. In 1985, when the survey was repeated, all the above-named species were gone, a change attributed to reduced water levels, loss of habitats, and the suppression of the flood-pulse cycle. In their place were a relatively small number of more-or-less ubiquitous introduced fishes, including largemouth bass and carp, that were able to cope with the new conditions.

The withdrawal of groundwater for irrigation is not the only postwar project that has left its mark on the ecological integrity of prairie rivers. Since the 1950s, engineers have also been busy constructing dams, diversions, impoundments, and other impediments to ensure that water is made available for human purposes. Across the entire Great Plains, with its thirsty cities and farms, there is scarcely a stream of any significance that has not suffered some such "improvement." For instance, the North Saskatchewan River is held back by two major impoundments before it even reaches the plains, pooling out to form Abraham Lake (in the mountains west of Red Deer) and the Brazeau Reservoir (southwest of Edmonton). The South Saskatchewan River is backed up by the Gardiner Dam (south of Saskatoon) at the head of the long noodlelike contours of Lake Diefenbaker. And this is to say nothing of the dozens of small reservoirs on dozens of tributaries that affect the movement of water through the Saskatchewan River system.

Then there's the Missouri. Now more accurately described as a string of reservoirs than as a river, the Upper Missouri is blocked by seven major dams, all but one of them built in the immediate postwar period. Working our way downstream from the river's headwaters, they are: Canyon Ferry (1952) and Fort Peck dams (1940) in Montana; Garrison (1953) in North Dakota; Oahe (1958), Big Bend (1963), and Fort Randall (1954) in South Dakota; and Gavins Point (1955), at Yankton, near the South Dakota/Nebraska border. Between Sioux City and St. Louis, by contrast, the lower Missouri has been intensively reengineered for a different use. Instead of being interrupted by dams, this "home stretch" of the river has been dredged, deepened, and banked to convert it into a fast-flowing, uninterrupted channel for barges. Together, these alterations have been reasonably effective at meeting their stated goals, whether navigation, power generation, recreation, or flood control. But they have also—to an extent that seems inevitable with hindsight but that was initially unforeseen—drastically altered the life of the Missouri.

The whole idea of damming rivers is to regulate seasonal variations in flow and make a chaotic natural system more predictable. Rather than permit spring meltwaters to rush out over the floodplain—by now a complex of cornfields, multilane freeways, and urban sprawl—engineers can hold back the flood behind massive barriers. With the exception of freak seasons like the springs of 1993 and 2019 (when rapid snowmelt and heavy rains combined to overwhelm the system and create havoc throughout much of the Missouri basin), the creative tumult of spring flooding is essentially over and done with. And it is not just the magnitude of the peak flows that is now under human control: their natural rhythm has also been altered. Under modern conditions, the volume of water that is permitted to pass through the sluicegates depends not on the logic of the seasons but on human demand. As a result, the more-or-less free-flowing stretches of the Missouri (between and below the dams) now typically experience their peaks not in spring but in winter, when the draw on the system is low, and fall to their lowest ebb during the growing season. This change completely confounds the flood pulse on which so many river-adapted organisms depend and makes the river a less livable environment for them.

Even changes that, at first glance, look like straightforward improvements often net a mixed bag of gains and losses. For instance, the Missouri, like many other prairie rivers, is naturally heavy with mud, the result of a long, slow journey through the dry and erodible soils of the grasslands. (They don't call it the Big Muddy for nothin'.) Also, like other prairie rivers, it is typically shallow—"a mile wide and an inch deep," as the settlers used to describe it. Warm, murky, and sullen, it is the antithesis of the proverbial babbling brook and, to some early observers, appeared to have few natural virtues. "Too thick to drink, too thin to plow," the old-timers grumbled. But if you hold this same water behind a dam, it will clear, as the load of dissolved sediments settles to the bottom. This is what happens in present-day reservoirs, creating perfect conditions—cold and sparkling clean—for sight-feeding game fish such as the native walleye (now far more common and widespread in the Missouri basin than in times past) and

LAKE STURGEON

Four members of the sturgeon family are found in the rivers and lakes of the Great Plains, and all of them are wonderfully odd. They include the pallid sturgeon (now one of the rarest fish in North America), the shovelnose sturgeon (a smaller, darker version of the pallid, so similar that the two frequently interbreed), and the paddlefish (perhaps the strangest of the lot, with its long, spatulate snout—a third of its entire length—which picks up electrical signals from the schools of plankton on which it feeds). All of these species are native to the Missouri/Mississippi drainage.

Only the final member of the family has a range that extends beyond the Missouri system and into the Saskatchewan basin. Found in both the North and South Saskatchewan rivers, the lake sturgeon looks more or less like the pallid—the same armored body, beady eyes, and whiskered chin—but can be considerably larger. The biggest lake sturgeons on record reached lengths of 8 feet (2.5 meters) and topped the scales at 275 pounds (125 kilograms). Gentle giants that they are, they subsist on a diet of fish eggs, insect larvae, and other small organisms that they scoop from the bottom. Slow grow-ing and slow to mature, they have been known to live for up to 150 years. Although the species came within a whisker of being lost from the North Saskatchewan River system, it appears to be coming back, thanks to improved water quality and a long-term ban on sport and commercial fishing.

Lake sturgeon

Pallid sturgeon

the introduced northern pike, smallmouth bass, black crappie, chinook salmon, and others. In all, about half the fish species in prairie rivers these days are exotics, and they tend to be doing very nicely.

But what of the poor unfortunates that *liked* the turbid water of the Missouri exactly the way it was? What about creatures like the pallid sturgeon? One of fifty-one species of native Missouri River fish that are known to be rare, uncommon, or in decline, the pallid sturgeon is among the unlucky few to merit inclusion on the federal list of endangered species. A member of an ancient order of cartilaginous fishes that has survived almost unchanged for some 70 million years, this is one weird-looking creature. It's big (up to 6 feet long and 100 pounds in weight, or 1.8 meters and 45 kilograms) and looks vaguely menacing, with its gray, armor-plated body; small, beady eyes; and flat, shovel-shaped proboscis. Four fleshy whiskers, or barbels, dangle under this protruding snout and finger the water in front of the recessed mouth. These waggling appendages are sensory organs that, through a combination of touch, taste, and electrical cues, permit the sturgeon to locate food in even the siltiest waters. (Pallid sturgeon eat small prey such as insect larvae and worms, which they vacuum up off the river bottom.) Interestingly, researchers in North Dakota have recently discovered similar adaptations in two native prairie minnows—the sicklefin chub and sturgeon chub—which, though weak-eyed, feel their way through the gloom by means of special sensory bumps located on their heads, fins, and bellies and inside their mouths. Thus, there appears to be a guild of prairie fishes that are specially equipped to cope with the special challenges of prairie rivers.

As the character of the habitat changes—as a silty, warm, seasonal river is transformed into a clear, cold, managed flow—these specializations lose their survival value. What is the point of being adapted to heavy siltation or to a seasonal flood-pulse or to a dynamic flow of habitats if many of these natural processes have been altered? This appears to be the fatal question for many native prairie species of fish, including the pallid sturgeon. A survivor of the Ice Age, it has not been able to adapt to the changes brought about by a half century of human ambition. Yet if evolution cannot undo the damage, perhaps people can. Although major dams are unlikely to be removed from the Missouri system any time soon, conservationists are working on ways to modify the obstructions so the fish can regain access to the remaining stretches of free-flowing current.

Even without human remediation, the pallid sturgeon, together with a number of other declining species, such as the western silvery minnow and the flathead, sicklefin, and sturgeon chubs, are managing to hang on in a few restricted reaches of the Missouri River. These include, as a prime example, a short run of water just east of the Montana/North Dakota border and west of Lake Sakakawea, near the confluence of the Missouri and the Yellowstone rivers. The last major river on the Great Plains without a major dam, the Yellowstone not only maintains its natural vitality—with May and June rises, fast- and slow-water habitats, natural temperatures and turbidity, and shifting banks—but also possesses enough energy to reinvigorate the smaller and more docile Missouri. The fact that the native community of fish respond to this wild influence, however locally, suggests that restoring sections of the river is a credible strategy. One of the places in which this kind of restoration is being attempted is the Big Muddy National Fish and Wildlife Refuge in the state of Missouri. From the early 1990s onward, stretches of the river have been released from the demands of navigation and allowed to flood their bottomlands, thereby creating new sandbars and side channels. In 1998, these efforts were rewarded by the discovery of baby pallid sturgeon, no larger than tadpoles, in one of the restored backwaters. The first sign of natural reproduction by the species in the lower Missouri in at least fifty years, these bewhiskered little creatures were a welcome reminder of the synergy between life and the living waters of the Great Plains.

CHAPTER 7

PRAIRIE WOODLANDS

And I saw that the sacred hoop of my people was one of many hoops that made one circle, wide as daylight and as starlight, and in the center grew one mighty flowering tree to shelter all the children.

BLACK ELK, *BLACK ELK SPEAKS,* 1932

I F YOU PAINT the prairie with broad strokes, all you see are grass and sky—a ground of green or khaki, a wash of blue. But scan the scene more closely, looking for shadow as well as for light, and you will see that the face of the grassland is etched with trees and shrubs. Tangles of woody vegetation are scrawled along the contours of the Great Plains, defining coulees, streambeds, and ridges and generally adding emphasis to a landscape that suddenly reveals itself in three dimensions. Far from being characterized by the absence of trees, these so-called treeless plains are increasingly a land of many small, scattered woods, each a world in itself, that add complexity to the grasslands.

It takes a prairie dweller to fully appreciate the way trees change the world. To sense how they cut the wind and scatter the light. To smell the rising odors of sap and moistness and rotting earth. To hear the silvery rustle of leaves and the bright chorus of birds—warblers, thrushes, vireos, wrens—that flit and flash through the branches. Every prairie grove is a woody oasis in a sea of grass, a refuge for the hundreds of species of insects, amphibians, reptiles, mammals, and birds that are adapted to the shelter of woodlands. Yet as much as we enjoy this abundance, we seldom stop to think about the woody beings that stand at the center of things. What are these impressive organisms that we call trees and shrubs? And what are they doing on the grasslands?

TO BE A TREE

THE UNDERLYING PRINCIPLE of woody plants is height. Through evolution, their strategy has been to hoist their leaves up into the light atop either a single, towering trunk (thereby producing the form classified as a tree) or a cluster of often-shorter, multiple stems (a shrub). The "technology" that makes this possible is an elaborately modified sugar, or organic polymer, called lignin. Produced by a layer of cells just under the bark, this remarkable material—durable, springy, and tough—is laid down toward the center of the stem or trunk, where it bonds with other plant tissues to form a supporting core of wood. Remarkably, this adaptation has been achieved independently by many different kinds of plants, including grasses (bamboo), legumes (caragana), and asters (sagebrush), as well as by all the varied tribes of conifers and hardwoods. In every case, the benefit has been the same: to gain an advantage over ground-hugging grasses and forbs by reaching up and intercepting the sunlight in midair.

But every adaptation brings its own difficulties. The closer a plant is to the heavens, the farther it is from Earth, and leaves, no matter how lofty, still must be fed by their roots. Like the photosynthetic organs of all plants large and small, tree leaves need a supply of water and mineral nutrients that can only come from the soil. Woody plants respond to this challenge by drawing groundwater up to the leaves through slender tubes in their phloem, or inner bark. Since the roots cannot push the fluid up, it has to be drawn from the top, a feat accomplished through evaporation. As the sun warms the surface of the leaves, water vapor is drawn through their pores, or stomata, and lost to the atmosphere. This loss creates a partial vacuum in the leaves that causes liquid to rise up the phloem tubes, each molecule glued to the next, as if it were in an improbably long drinking straw. During the growing season, a large deciduous tree can suck up as much as 100 gallons (375 liters) of water in a day, just to keep its vital juices flowing.

But if the roots cannot find enough water to satisfy this hectic demand, the slender threads of liquid in the phloem will break, or cavitate—often with an audible pop—and circulation to part or all of the upper tree will be cut off. Death, or die-back, follows. This is exactly what happened, on a large scale, during the drought of the Dirty Thirties when, to cite one well-documented case, between 50 and 60 percent of all trees in Kansas and Nebraska succumbed to water stress, and most of the survivors were left stunted and partially blackened. In northern Texas and Oklahoma, where the death toll was even worse, as few as two or three trees in ten survived the decade of drought.

For a tree on the sunbaked prairies, survival is a question of supply and demand. On the supply side, trees reach out for water by producing ambitious systems of roots, which are often almost as massive as the plant's aboveground growth. Contrary to a common misconception, most tree roots are relatively shallow—no more than a few yards, or meters, deep—and they rarely tap into the permanent moisture of the water table. (Even roots can drown if they're held under water and deprived of oxygen for too long.) Instead, the plants push out laterally through the rain-watered upper layers of the soil until their rootstocks extend well beyond the circumference of the branches. Green ash, for example—an eastern woodland species that occurs midway across the plains—puts out roots to a distance of some 50 feet (15 meters), but rarely pushes them more than 3 feet (roughly a meter) deep. The more arid the habitat and the more drought-resistant the tree, the more far-reaching the roots tend to be. Thus, the gnarly little bur oak—one of the most drought-hardy species of trees on the tall-grass prairie and one of the few that carried on merrily during the thirties' drought—typically produces roots that are as deep as the tree is tall and more than twice as extensive. To be specific: a forty-three-year-old bur oak that had struggled up to a height of 20 feet (6 meters) was found to have pushed its roots out laterally over a distance of 41 feet (12.5 meters).

On the demand side, trees reduce the rate at which water vaporizes and streams away from the surface of their leaves. Deciduous trees, for example, take the radical approach of shedding their leaves in autumn, entering near dormancy, and then bursting forth with new vigor at the time of the spring rains. Conifers have reduced the surface of their leaves to the barest minimum, thereby cutting their water through-put, slowing down their metabolism, and allowing them to photosynthesize sluggishly throughout the year. In addition, the leaves and needles of many species of trees are covered with a waxy, waterproof coating, or cuticle, that helps keep moisture from leaking out, except through the portholes of the stomata. And even these pinprick openings can be regu-lated to a certain degree—opened when the air is moist and cool or shut when it's dry and hot—to minimize the risk of drying out.

Some species of trees, and even some geographical pop-ulations within a given species, are better than others at turning off the tap. For example, trembling aspens that grow on semiarid grasslands have a greater ability to adjust their stomata and reduce their water loss than do aspens from other, less-challenging environments. But despite these refinements, a typical tree is still a large and thirsty organism that often finds itself tested by the supply-demand equations of the open plains. This helps explain why prairie woodlands are often tucked into folds of the land, close to the water table and protected from sun and wind, where the circumstances are most forgiving.

WHERE THE TREES ARE

UNTIL RECENT TIMES (before agricultural settlement began), the limits of life on the Great Plains were largely set by

AMAZING ASPENS

Of all the trees that grow on the prairies, none can bounce back from disaster more quickly than the trembling, or quaking, aspen. ("Stand still for half an hour," one biologist was heard to complain, "and those darned things grow right up your pant legs.") The key to their vitality lies in their roots, which have a grasslike ability to put up new shoots. The root system is typically shallow but extensive, with lateral runners that can reach out 100 feet (30 meters) and sinkers that descend 6 to 10 feet (2 to 3 meters). The runners are studded with growth points, as many as thirty to an inch, each capable of producing a new stem and, potentially, a new aspen. Such growth occurs in the first years after a fire or drought, as saplings shoot up by the thousands and then are thinned out through the survival of the fittest.

Although each stem eventually develops a more-or-less separate system of roots, there is a real sense in which a grove of aspen is a single plant. Because all the trees in the stand have risen from the same parent stock, they carry identical DNA and share such characteristics as protein content, sprouting rate, and insect resistance. Each bluff also responds to the seasons at its own genetically programmed pace, so related trees tend to green up and drop their leaves in synchrony. Since aspens are dioecious (with separate male and female plants), all the members of a particular clone are usually also of the same sex. The largest known organism alive today is a stand of 47,000 male stems in south-central Utah.

Trembling aspen

Aspen twig
with bud scars

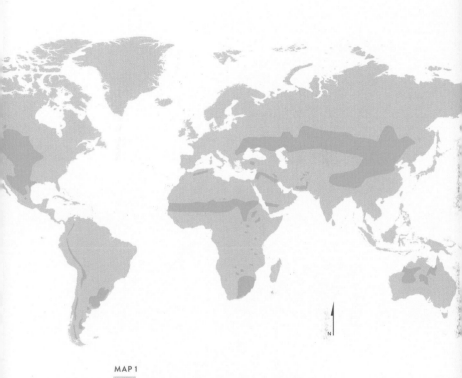

MAP 1

TEMPERATE GRASSLANDS OF THE WORLD

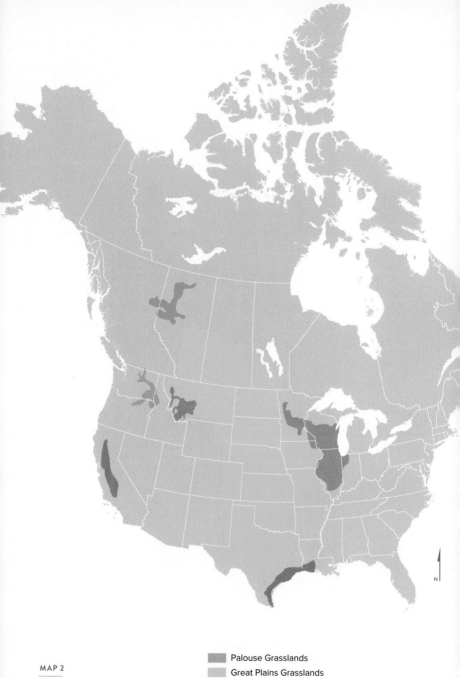

MAP 2

**TEMPERATE GRASSLANDS
AND SAVANNAS OF CANADA
AND THE UNITED STATES**

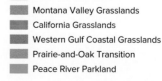

Palouse Grasslands
Great Plains Grasslands
Montana Valley Grasslands
California Grasslands
Western Gulf Coastal Grasslands
Prairie-and-Oak Transition
Peace River Parkland

0   500
         kilometers

0   500
                 miles

N

Eagle Hills

Pasquia Hills

Red Deer Badlands

Great Sandhills

Duck Mt.

Manitoba Escarpment

Porcupine Hills

Riding Mt.

Sweet Grass Hills

Cypress Hills

Wood Mt.

Pembina Hills

Bears Paw Mts.

Turtle Mt.

Highwood Mts.

Little Rocky Mts.

Missouri Coteau

Pembina Escarpment

Red River Lowlands

Judith Mts.

Big Snowy Mts.

Little Missouri Badlands

Crazy Mts.

Bighorn Mts.

Bear Lodge Mts.

Black Hills

Pine Ridge Escarpment

MAP 3

GEOGRAPHY OF THE GREAT PLAINS

Nebraska Sandhills

Loess Hills

Alberta Plain

Saskatchewan Plain/ Glaciated Central Lowlands

Missouri Plateau

High Plains

Colorado Piedmont

Plains Border

Osage Plains

Raton Section

Pecos Valley

Edwards Plateau

Great Bend Sand Prairie

Caprock Escarpment

Mescalero Escarpment

Llano Estacado

Badlands

Sandy area

Mountain

Hills

Escarpment

Maximum extent of all glacial advances in the Pleistocene epoch

Mescalero Sands

500 kilometers

500 miles

Balcones Escarpment

N

MAP 4

**ECOREGIONS OF THE
GREAT PLAINS**

Short-to-Mixed Grasslands
Northwestern Short/Mixed Grasslands
Southern Short Grasslands

Mixed Grasslands
Northern Mixed Grasslands
Foothills Grasslands
Nebraska Sandhills Mixed Grasslands
Southern Mixed Grasslands

Tall Grasslands
Northern Tall Grasslands
Central Tall Grasslands
Flint Hills Tall Grasslands
Texas Blackland Prairies

Woodlands and Savannas
Cross Timbers Forest
Black Hills Coniferous Forest
Aspen Parkland
Southern Prairie-and-Oak Transition
Edwards Plateau Savannas

0   300
      kilometers

0   300
                    miles

N

Arctic

Pacific

Desert

Gulf

0 — 500 kilometers

0 — 500 miles

N

MAP 5

MAJOR AIR MASSES AFFECTING THE GREAT PLAINS

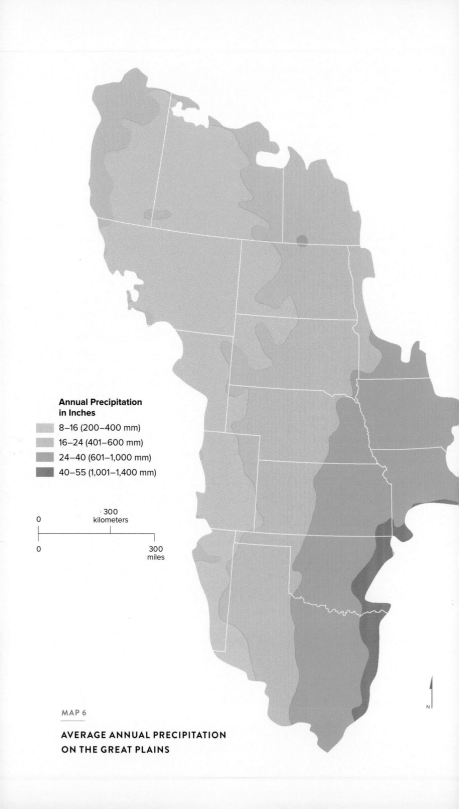

Annual Precipitation
in Inches

8–16 (200–400 mm)
16–24 (401–600 mm)
24–40 (601–1,000 mm)
40–55 (1,001–1,400 mm)

0   300
kilometers

0   300
miles

MAP 6

AVERAGE ANNUAL PRECIPITATION
ON THE GREAT PLAINS

N

MAP 7

SOILS OF THE GREAT PLAINS

Alfisols
Soils with a surface zone of silicate clay

Moist to wet, warm soils

Dry, warm soils

Aridisols
Soils of arid environments with subsoil development

Dry, warm soils

Entisols
Soils with little morphological development

Dry, warm soils

Dry, warm, sandy soils

Mollisols
Grassland soils

Moist to wet, warm soils

Dry, warm soils

Moist to wet, cool soils

Vertisols
Clayey soils with high shrink/swell capacity

Warm soils of all soil suborders

Ultisols
Acidic soils with clay-rich subsoils

Warm soils of all soil suborders

Inceptisols
Moist soils with little subsoil development

Warm and cool soils

MAP 8

**GRASSLAND COVER ON
THE GREAT PLAINS**

Pixels (data points) of
100 percent grassy cover,
from satellite imagery

Pixels of mixed grassy
cover and cropland,
but with a predominance
of grass

300
kilometers

0

0   300
                         miles

N

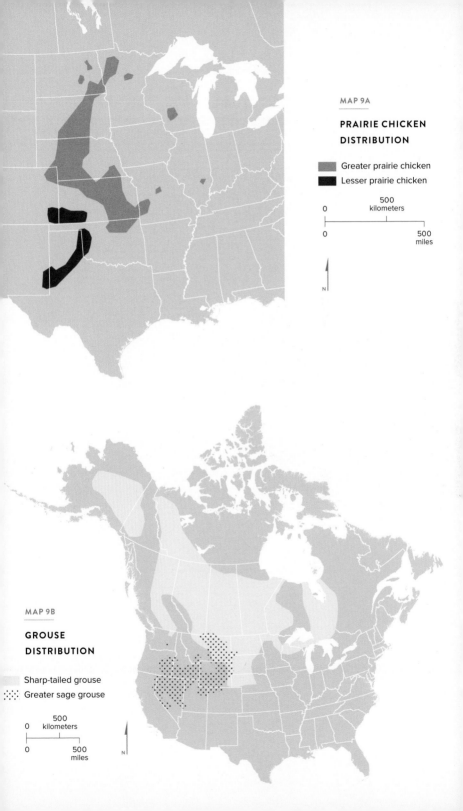

PRAIRIE CHICKEN DISTRIBUTION

Greater prairie chicken

Lesser prairie chicken

500
kilometers
0

0   500
miles

N

GROUSE DISTRIBUTION

Sharp-tailed grouse

Greater sage grouse

500
kilometers
0

0   500
miles

N

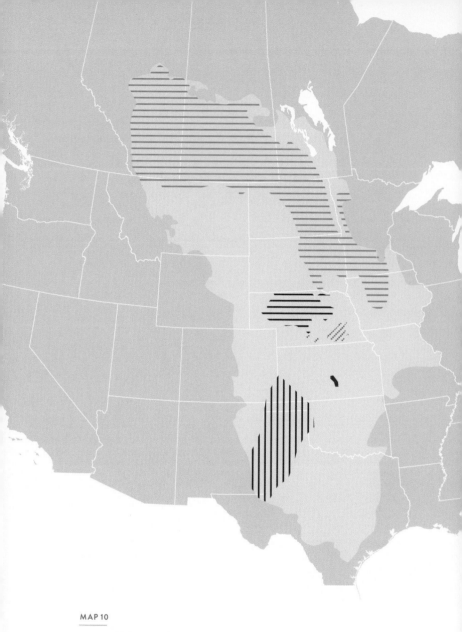

MAP 10

WETLAND REGIONS
OF THE GREAT PLAINS

Prairie Pothole Region

Nebraska Sandhills Wetlands

Rainwater Basin

Central Kansas Wetlands

Playa Region

500
kilometers

0

0

500
miles

N

MAP 11

**RIVER SYSTEMS OF
THE GREAT PLAINS**

— Saskatchewan River System

— Missouri/Mississippi River System

0
500
kilometers

0
500
miles

N

MAP 12

**WOODLANDS AND SAVANNAS
OF THE GREAT PLAINS REGION**

Aspen Parkland

Edwards Plateau Savannas

Southern Prairie-and-Oak Transition

Northern Prairie-and-Oak Transition

Black Hills Coniferous Forest

Cross Timbers Forest

500
kilometers

0

0

500
miles

N

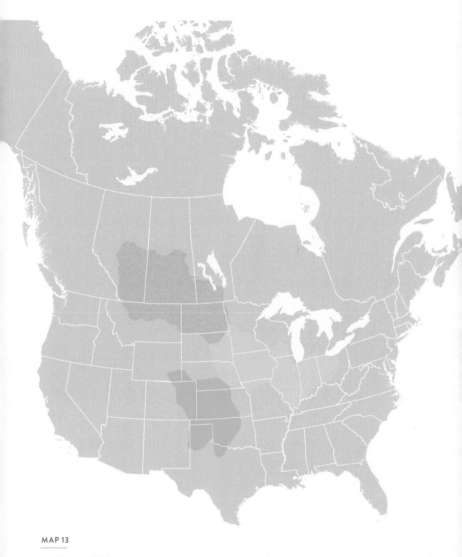

MAP 13

**MAJOR CROPS
OF THE GREAT PLAINS**

Spring Wheat Belt
Winter Wheat Belt
Corn Belt

0   500
       kilometers

0   500
             miles

N

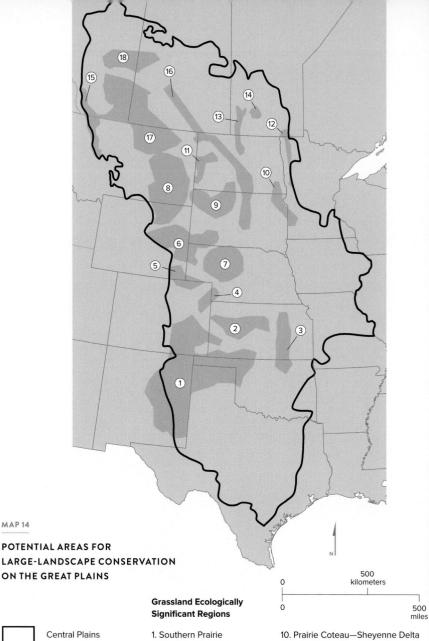

MAP 14

**POTENTIAL AREAS FOR
LARGE-LANDSCAPE CONSERVATION
ON THE GREAT PLAINS**

Central Plains

Grassland Ecologically
Significant Regions

**Grassland Ecologically
Significant Regions**

1. Southern Prairie	10. Prairie Coteau—Sheyenne Delta
2. Smoky Hills	11. Little Missouri
3. Flint Hills	12. Agassiz Tallgrass
4. Arickaree	13. Manitoba Mixed-Grass—Towner Sar
5. Pawnee Grasslands	14. Carberry Sandhills
6. Thunder Basin—Conata	15. Rocky Mountain Front
7. Sandhills	16. Missouri Coteau
8. Powder River—Big Open	17. Medicine Line Mixed-Grass
9. Cheyenne Prairie	18. Northern Fescue

500
kilometers

0

0   500
                                    miles

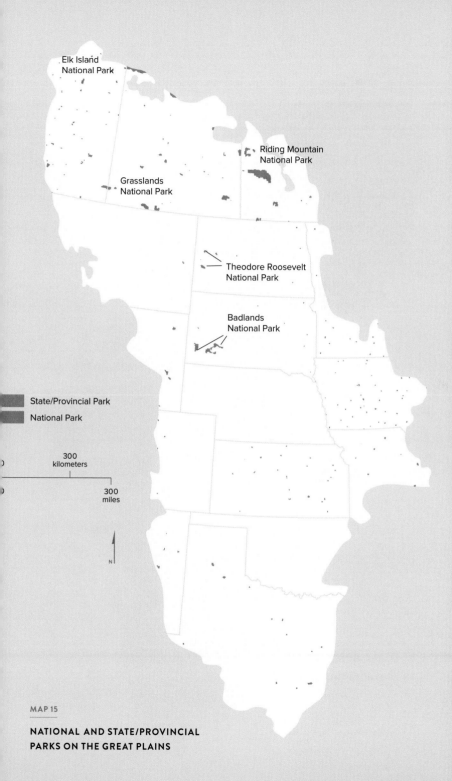

Elk Island
National Park

Riding Mountain
National Park

Grasslands
National Park

Theodore Roosevelt
National Park

Badlands
National Park

State/Provincial Park

National Park

300
kilometers

300
miles

N

MAP 15

**NATIONAL AND STATE/PROVINCIAL
PARKS ON THE GREAT PLAINS**

MAP 16

THE GREAT PLAINS

0 ⊢──────┤ 300
kilometers

0 ⊢──────┤ 300
kilometers

N

weather. The invisible contours of the climate—especially large-scale gradients of temperature, precipitation, evaporative demand, and wind—were mapped onto the landscape as visible differences in vegetation. The same west-to-east trend in precipitation that governed the transition from short-grass through mixed-grass to tall-grass prairie had a parallel influence on the height, density, and vigor of woody vegetation. In the dry short-grass ranges of the western plains, for instance, the brush tended to be scrubby, sparse, and tough, dominated by slow-growing, drought-hardy species such as sagebrush and juniper. And in the extreme southwest, where low precipitation was compounded by searing heat, the land was invaded by shrubs from the desert chaparral, including honey mesquite, creosote bush, and the ferociously barbed cat's claw mimosa. Even with this intrusion, as little as 3 percent of the western plains was wooded. *See Map 12: Woodlands and Savannas of the Great Plains Region.*

Moving eastward into the mixed-grass zone, the shrubbery gradually became thicker and more lush, filling the draws with bushes such as chokecherries, hawthorns, and wild roses. Even the grassy uplands were studded with clumps of snowberry and silverberry (also known, respectively, as wolfberry and wolf willow). By the same token, the banks and bottomlands of the major rivers, which had been covered with grass farther west, became shrubby and then heavily treed as they cut east through the mixed grasslands. By the time they reached the tall-grass prairie, these streamside, or

American elm

riparian, woodlands had taken on the character of full-fledged forests, with willows and cottonwoods along the shore, hardwoods such as green ash and American elm (with their understory of shrubs) on the lower slopes, and bur oak at the top of the banks, where they faced off against the blue-stems and other grasses. These serpentine strips of riparian woodlands contributed to the estimated 15 percent of the tall-grass region that, under natural conditions, was covered by woody plants.

If one left the river valleys and headed still farther east—beyond the basins of the Red and Missouri rivers, toward the eastern hardwood forests—the country went through a passage of ecological uncertainty, in which it was never quite clear if the trees or the grasses were in ascendancy. Still tracking along the same west-to-east increase in precipitation, the tall grasses were increasingly invaded by stands dominated by hickories and oaks until these groves eventually coalesced into a dense deciduous forest. This grassland-to-forest transition occupied a broad arc that swept south from Minnesota to Missouri and then (under the sway of moist weather rolling in from the gulf) swung to the southwest, following the curve of the Ozarks in the direction of Texas. There, at the southern terminus of the transitional zone, lay the Cross Timbers Forest, a picturesque savanna of weather-beaten oaks and junipers interspersed with glades of grass that once sprawled over some 20 million acres (8 million hectares) in southwestern Kansas, Oklahoma, and central Texas. Although most of the trees were logged off in the early 1900s, significant fragments of the primordial Cross Timbers remain, including gnarled cedars that date back to the sixteenth century.

OLD-GROWTH SPECIALISTS

The phrase "old-growth forest" conjures up images of towering redwoods and firs in the gloom of the Pacific rain forest. We're much less likely to think of scraggly old juniper or oak trees, out on the burning plains, that have taken four hundred years to reach the height of a two-story building. Yet stunted and bent as they are, these venerable prairie warriors deserve the respect due to age and to increasing rarity.

Take, for example, the twisted trees of the Texas Hill Country. Originally an open savanna of mixed woods interspersed with meadows of grass, the region was heavily logged in the late 1800s. Then, thanks to fire suppression, woody plants pushed back in, often as dense, monotypic, pollen-spewing stands of the native "cedar," or Ashe juniper. Despised by ranchers, developers, and allergy sufferers alike, Ashe juniper was attacked with a vengeance.

Yet Ashe juniper has its uses, especially to a bird called the golden-cheeked warbler. A close relative of the widespread yellow warbler, this species can be recognized by appearance (dark body, yellow face with black eye stripe), by song ("bzzzz layzee dayzee"), and by location (it nests only in central Texas). Most at home in extensive tracts of mixed woodlands, the golden-cheek has a specific requirement for mature Ashe junipers, aged twenty or thirty and up, that have developed shredding bark. The female warbler uses these fibers, bound with spiders' silk, to build her soft, gray, perfectly camouflaged nest. As shaggy old junipers have become scarce, so have the golden-cheeks. Predation by blue jays from encroaching urban forests and parasitism by cowbirds compound the problems, which do not offer easy solutions.

Golden-cheeked warbler

A comparable transition zone, or ecotone, is also found on the prairies' northern shore, where mixed grasslands lap up against the boreal forest. Just as the west-to-east increase in precipitation encourages the growth of trees, so the south-to-north decline in average temperatures reduces the evaporative demand and creates conditions that are hospitable to forests. Somewhere around 52 degrees north latitude (about the latitude of Saskatoon), the climate strikes a balance between the requirements of certain midheight grasses—rough fescue, for example, and western porcupine grass—and those of various trees and shrubs, notably, trembling aspens. The result is a pleasantly varied landscape of rolling, grassy meadows and contoured groves, or bluffs, that is known as Aspen Parkland. Extending perhaps 60 miles (100 kilometers) from north to south, the parklands stretch all the way across the Prairie provinces, from the foothills of central Alberta to the Canadian shield, and south through Manitoba into northwestern Minnesota. There the broken stands of aspens shade imperceptibly into the savannas of hickory and oak that extend, with remarkably few interruptions, south to the Cross Timbers.

Thus, for the most part, prairie woodlands sort themselves along the gradient of the four cardinal directions, north, south, west, and east, in accordance with their ability to withstand water stress. On a finer scale, they tuck themselves into the prairies wherever the lay of the land either improves the supply of water to the roots or (through shelter and shade) reduces the evaporative demand on the foliage. That is why, for example, the north-facing slope of a coulee is often densely tangled with brush, while the exposed southern slope supports nothing but a carpet of forbs and grasses.

ICE AGE RELICS

THE TENSIONS BETWEEN supply and demand also help to account for another little-noticed feature of the prairie landscape. You'll be scooting across the flatlands when suddenly the way ahead is obstructed by an abrupt rise. As the road climbs the escarpment, the view tilts toward the light and a phalanx of bristling conifers appears against the sky. Why, you may well wonder, would an evergreen forest sprout on the top of a rocky ridge, smack dab in the middle of the prairies? Yet this phenomenon is repeated in dozens of places across the Great Plains, from the white-spruce forests of the Cypress and Black hills to the ponderosa pines of Pine Ridge to the junipers of the Caprock Escarpment, among many others.

These mysterious scarp woodlands are relics from the past. At the end of the last Ice Age, the climate was cool and damp, and a dense coniferous forest stretched across the breadth of the continent. A dark mantle of spruce extended from what are now the Canadian prairies south through the Dakotas into the central states, while pine woods appear to have flourished on the southern plains. But as the chill of the glaciation gradually lifted, the climate eased into a drier and warmer phase, marked by more frequent droughts, and the boreal forest was forced to retreat to the north. Within remarkably short order—a blink of the cosmic eye—the forests of the Great Plains either surrendered directly to grasses or else gave way first to deciduous trees and then to prairie.

By the time the transformation was over, coniferous forest could only be found on the crowns of the tallest breaks and ridges. High enough to catch the rain and snow, and cooler

FOSSIL FORESTS

Relics of the late Ice Age forest are strewn across the Great Plains, rather like flotsam from an ecological shipwreck. In the Sweet Grass Hills of Montana, for example, stands of Rocky Mountain Douglas fir are found alongside hybrid (white × Engelmann) spruce, growing just where they were left stranded thousands of years ago. The main populations of all three species are now far to the north and west, in the mountains and the boreal forest. Similarly, scientists working in Wyoming and Nebraska have discovered relict stands of hybrid poplars (cottonwood × balsam) in parts of the Niobrara Valley, where the two species had not been in contact since the late Pleistocene.

The distribution of these living fossils has helped researchers to get a sense of the extent of the prehistoric forest. For instance, the fact that there are moose in the Turtle Mountain/Pembina Hills region of Manitoba and North Dakota has been taken as evidence that the northern plains were once covered by extensive woodlands. By the same token, the flora and fauna of the Black Hills, with their disjunct populations of white spruce, paper birch, moose, black bears, and other boreal species, speak of a time long past when coniferous forest flowed out of the Rocky Mountains and across the Great Plains.

than the grasslands below, these uplands created a microclimate in which the trees could retain a toehold. The thin
mineral soils of the ridges—more suitable for conifers than
for grasses or other plants—probably also gave the trees an
advantage. But the factor that made the biggest different to
the scarp forests was their top-o'-the-world location. Histori-
cally speaking, the greatest threat to prairie woodlands, apart
from prolonged drought, was the fierce heat of grass fires.
Where better to find refuge than atop a natural firebreak, an
outcropping of safety in a world of flame? The present distribution of scarp woodlands therefore probably represents the
limits of wildfires past, a kind of high-tide line beyond which
the danger did not pass.

TREES ON THE MOVE

FOR THE WOODY plants that grew down on the grasslands,
by contrast, there was no chance of escape. Despite a variety of adaptations for withstanding occasional burns—corky,
fire-resistant bark (bur oak); seeds that are stimulated by heat
(ponderosa pine); roots that put out new shoots to compensate for fire damage (many poplars and oaks)—few species
of trees can survive frequent, intense fires. Grasses, by contrast, are basically born to burn. Not only do they produce a
tinder-dry thatch of dead foliage that lights with the slightest
spark, but they are also equipped to rise from their own ashes.
The buds, or meristems, from which they put up new growth
are tucked down at the surface of the ground, where they are
protected from serious harm. But woody plants, which are
inclined to reach for the light, hold their buds on the tips of

their branches, where they are exposed to the flames and are sorely vulnerable to fire damage.

In the days of the buffalo prairie, frequent grass fires conspired with severe drought and occasional intense grazing to limit the spread of woody plants. In the sometimes-lengthy interludes between die-backs, trees and shrubs were often able to take advantage of cool, moist weather to extend their reach onto the grassy plains. But no sooner were the trees established than some random bolt of lightning would set the prairies aflame, killing shrubs and most trees except in the humid coulees and river valleys. On the tall-grass prairies, in particular, where catastrophic droughts were relatively infrequent, trees and shrubs might have taken over completely if it hadn't been for the erratic but inevitable return of lightning.

Prairie fires were also set by Indigenous peoples, who used burning as a tool to hold back the brush and maintain the grasslands as pasture for bison. But with the end of the buffalo ecosystem and the introduction of agriculture, prairie fires were suppressed and trees began to make a slow but steady advance. This expansion of woody vegetation has now been documented in almost every ecoregion across the Great Plains, from the Aspen Parkland of central Alberta (where a 60 percent increase in the area of brush was noted between 1907 and 1966) to the Flint Hills of Kansas (where patches of woodland protected from burning have enlarged by 250 percent since the 1850s) to the dry grasslands of west Texas (where the area covered by mesquite has more than tripled locally in just over a century). At the same time, various species of junipers are intruding on western rangelands, ponderosa pines have expanded from the Niobrara River valley into the Nebraska Sandhills, and dense stands of eastern red cedar dominate the once-grassy slopes of Iowa's Loess Hills. These examples could be multiplied many times over.

Extrapolating from local experience to calculate the effects on the Great Plains as a whole has not proven to be simple. Estimated area of presettlement woodlands *minus* acreage lost to clearing and other development *plus* postsettlement expansion of woody growth *equals* the sum of many unknowns. Still, the available evidence strongly suggests that the Great Plains region is now more heavily wooded than it was two hundred years ago. The big, unanswered question is, why? Some researchers have suggested that the changes began back in the days of the cattle barons of the nineteenth century, when the prairie

Eastern red cedar

was badly damaged by overgrazing. As the hold of the grass was weakened, patches of bare soil were exposed, creating openings where woody plants could grow. Once the brush was established, it was able to expand. Or perhaps the cattle not only broke the sod but also sowed the seeds by ingesting the fruits of trees and shrubs (notably, honey mesquite) and distributing them in their droppings. Alternatively, it is possible that global, rather than local, forces have been at work. Recent evidence suggests that the rising concentrations of carbon dioxide in the atmosphere, due to the burning of fossil fuels, may be encouraging the growth and expansion of trees and shrubs by giving them a physiological advantage over nonwoody plants. But whatever other factors have been involved, the suppression of the natural fire regime has certainly played a central role in encouraging the intrusion of woody plants into the grasslands.

IN THE RIPARIAN ZONE

THE TRANSFORMATION OF grasslands into woodlands is far from trivial. It represents the difference between ground squirrels and tree squirrels, meadowlarks and robins, sharp-tails and ruffed grouse, between the stripped-down economy of the drylands and the richer options of a more sheltered, and sheltering, wooded environment. Of all the habitats on the Great Plains, none is more biologically productive—or more subject to disturbance—than the sinuous stands of trees and shrubs that line the major creeks and rivers. Although these riparian forests occupy only about 1 percent of the region, they provide living space for somewhere between 60 and 90 percent of the species of vertebrates (reptiles, amphibians, mammals, and birds) that occur on the prairies. For example, researchers working in northeastern Colorado were amazed to discover that 82 percent of all the local birds could be found down by the river. This finding helps to account for the delight of walking along a prairie river on an early morning in spring, as a dozen different bird songs shimmer above the canopy. For anyone who has experienced this pleasure, it will come as no surprise to learn that, in both numbers of species and numbers of individuals, these wood-lands are home to some of the richest avian communities in North America.

But the resources of the riparian zone do more than pro-vide habitat for an abundance of wild animals and birds. Over the years, they have also proven highly attractive to humans, from the seminomadic people of the Plains Wood-land tradition, who gardened along creeks in Colorado some two thousand years ago, to the village-dwelling Mandans and Hidatsas, who, as recently as the 1880s, were growing corn, beans, and squash in plots along the Missouri River and its tributaries. With the intrusion of industrial society, the demands on the riparian zone have grown ever more intense, as we look to the river valleys for farmland, rangeland, roadways, reservoirs, and river-view subdivisions. When these localized impacts are overlain on the global effects of fire suppression and climate change, the riparian community can be expected to respond in conflicted and complex ways.

Even the most innocent changes can trigger profound effects. For instance, what could look less stressful than a herd of cud-chewing cattle bedded on a riverbank, watching the water slip quietly past? Yet wherever they are present, cattle present a major threat to the health of riparian ecosystems. Unlike the native-born bison, which were well adapted to life on the dry plains, cattle evolved in more temperate environments and are drawn to treed valleys in search of water, shade, and forage. Although moderate grazing, by a small number of cattle for a short period of time, may not cause noticeable harm, heavy grazing inevitably leaves deep scars. Too many hooves in too little space soon pound the place to death, as the animals foul the water, destabilize the banks, and trample or chew the upcoming crop of woody plants. Traditionally written off as sacrifice areas by the cattle industry, riparian woodlands have only recently been appreciated as a rich resource for wild animals and plants. Thanks to this new awareness, range managers are promoting the use of artificial water points, streamside fencing, and other strategies to protect these "hot spots" in the living landscape.

Unfortunately, damage caused by other means is often more difficult to correct. In the central United States, for example, corridors of hardwood forest once spread across the broad floodplains, or bottomlands, of rivers and creeks, including (among others) the Milk, the Marias, the Wildhorse, and the mighty Missouri. The riparian forests of the Missouri River, for example, opened out in the Dakotas and, like the river itself, grew broader and more majestic as they flowed south, expanding from a width of about half a mile (800 meters) in the north to a span of almost 19 miles (30 kilometers) as they neared the river's mouth. Today, this irresistibly flat and fertile plain has been mostly converted to farms. Between the Nebraska state line and the Mississippi River, more than 80 percent of the lands that were forested before settlement are now planted to crops like corn and sorghum.

Farther upstream in the Dakotas, the outcome has been similar, though the means have been more complex. Here, too, forests have been lost to clearing: for instance, almost 60 percent of the native woodlands between the Dakotas' Garrison and Oahe dams are now under cultivation. As for the remainder, most of the once-wooded floodplain now lies under the waters of lakes Sakakawea and Oahe and other impoundments. Although the shores of the reservoirs are fringed with patches of brush, these fragmented woodlands are neither as extensive nor as diverse—nor as rich in species of birds—as the original forests were. And even in places where remnants of the native forest have survived (for instance, along undammed stretches of the river west of Williston, north of Bismarck, and east of Yankton), things are not looking good. The problems first surfaced in the mid-1970s, when an analysis of cores from trees in the Garrison-to-Bismarck stretch of the river, downstream from the Garrison Dam, showed that they were not growing as quickly as they had twenty years earlier, before the dam was constructed. Somehow or other, the impoundment of the river was affecting the health of riparian ecosystems over distances of as much as 60 miles (100 kilometers).

How could a stationary wall of earth and concrete exert this kind of influence? The answer was that it was doing what it had been designed to do: controlling the flow of the river and preventing flooding. By depriving the riparian forests of a well-timed influx of water from spring floods, the dam had cast a shadow over them. In response, all the major species of trees in the forests, including green ash, box elder (or Manitoba maple), and American elm, were suffering from reduced rates of growth. And then there were the cottonwoods.

The dominant trees of the riparian zone across the prairies—and often the *only* species on the arid plains of Alberta, Montana, and other points west—cottonwoods are dependent on the energy and flux of flood for their very survival. Each spring, a typical grove of cottonwoods pumps out billions of cottony, wind-borne seeds, most of which end up drifting forlornly around the country. Only a tiny fraction find the conditions that they need for germination and growth: bare soil, full sun, and plentiful moisture. Where better to meet these requirements than in the wake of a spring flood? As the river recedes from the floodplain, a terrain of newly deposited sandbars and clean-scoured banks is exposed, creating ideal conditions for cottonwood seeds to sprout. (That is why cottonwoods often grow in bands and arcs of same-age trees, each representing a historical moment of opportunity.) But if spring floods are constrained and seed beds are not produced, the trees lose their only chance at reproduction for an entire year.

As Wallace Stegner once pointed out, "western history is a series of lessons in consequences," and the consequences for cottonwoods have, for the most part, not been good. Because their rate of reproduction is reduced, the trees can't compensate for natural mortality, and many populations are collapsing. (Cottonwoods are old by age thirty, and few last longer than a century.) The surviving stands of cottonwoods along the Missouri, Bighorn, Milk, South Saskatchewan, and several other rivers are all dying off, and there are few ambitious young saplings to take their places. As early as 1981, for instance, just a couple of decades after the Waterton and St. Mary rivers of southeastern Alberta were dammed, the number of cottonwoods along their banks had declined by 23 percent and 48 percent, respectively. By contrast, along the neighboring Belly River, which was not restrained by dams, the cottonwoods continued to reproduce and sustain themselves.

Where the trees are dying off, they are being replaced, in the west, by grassy banks or, farther east, by species of trees that are not dependent on seasonal flooding. These include both native trees such as ash, maple, and elm and invasive, introduced "weeds" such as tamarisk (salt cedar) and Russian olive. But even if other trees do grow up to fill the gap, the absence of cottonwoods is a loss to the roughly one-third of prairie birds that rely on them for food and shelter. Because (like other members of the poplar tribe) they support an exceptionally rich fauna of insects, they are an especially valuable resource for insect eaters such as warblers and woodpeckers. And because of their propensity for heart rot (another poplar attribute), they are an important source of nesting habitat for woodpeckers, chickadees, tree-nesting ducks, and other cavity nesters. Cottonwoods also produce fat, juicy leaf buds that serve as a vital wintertime food for fat, juicy browsing birds such as sharp-tailed and ruffed grouse, which teeter improbably in the upper branches.

Bufflehead at nest

THE CASE OF BRAIDED RIVERS

INEVITABLY, ON A planet that always has a surprise or two up its sleeve, there is a twist to this story—a special case in which the same suite of human actions has brought about an equal but opposite effect. Along certain rivers in the west-central plains, notably, the Arkansas in eastern Colorado and the Platte system of Wyoming, Colorado, and Nebraska, the curtailment of spring floods has had the unexpected effect of *promoting* the growth of cottonwood forests. To reproduce successfully, cottonwoods need bare, flood-scoured seed beds for germination. In succeeding years, as young saplings, they also need a respite from flooding. A so-called meandering river like the Missouri throws up banks and sandbars as it carves and recarves its course, thereby creating slightly raised beds on which some of the saplings, some of the time, find safe haven. But a shallow, braided river like the Platte runs and ripples across its streambed and, under natural conditions, is likely to scour out any seedlings in its channel. Even if a cottonwood seed does manage to put down roots, it will probably be swept away by the river long before it can reach adulthood.

Or at least this is how things used to work, before the dams went in. According to land surveys from the middle to late 1800s, the banks of all three Platte rivers—the north and south forks and the Platte River itself—were originally covered in grass, with only scattered clumps of brush and little or no cottonwood forest. But by the 1950s, the waters of the Platte system had been impounded behind several dozen dams, and peak flows were dramatically reduced. In response to these changes, each of the three rivers dwindled into progressively narrower channels, thereby abandoning much of their streambeds and exposing a series of beaches, or natural levees, on which trees and bushes could become established. By the time the channels stabilized in the late 1960s, the streams had been reduced to between one-fifth and one-twentieth of their historic widths, and their once-open beds were scrawled with willows and lined with a gallery of towering cottonwoods.

Once set in motion, the wheels of ecological change have just kept on rolling. Because the rivers are now in balance with their flows and are no longer narrowing, they have effectively stopped producing nursery sites for cottonwood trees. It follows that these forests will soon suffer the same fate as the cottonwoods everywhere else: they will die off and be replaced by mixed hardwood forests. This coming transition is already signaled by the advance of Russian olive throughout the Platte drainage, a silvery foreshadowing of future change. But despite these continuing adjustments, the forests that have sprung up along the Platte system appear to be here to stay, at least for as long as the rivers are constrained.

GO WEST, YOUNG BIRD

FROM A BIRD lover's point of view, it may be hard to understand why anyone would object to the expansion of woodlands along the Platte or anywhere else on the Great Plains. The more trees have invaded the grasslands over the last hundred-odd years, the more species of birds have moved in, either as summer residents or as visiting migrants. No matter how the expansion of woody growth was triggered—whether through overgrazing or climate change or the suppression of fire and flood—the result has been a diversification of the local environment. The addition of woody vegetation brings a complex vertical dimension to the living world that extends from the ground up, through the shrub layer, into the sheltered midcanopy and the treetops. At every level, trees and shrubs provide habitats to a particular community, or guild, of breeding birds that is adapted to specific conditions of light, humidity, and shelter. In addition, woodlands also offer a range of options along the radius of each wooded stand, from the shadowy stillness of the interior to the bright and breezy margins. For an ovenbird, a secluded forest interior. For a catbird, a shrubby edge. For an oriole, the airy heights of the upper branches.

Life responds to opportunity, and the expansion of woody vegetation across the plains has opened up new opportunities for dozens of species of birds that were seldom, if ever, seen here by early settlers. If the first generations of prairie naturalists could peer down from heaven, binoculars in hand, they would be astonished by the flood of recent arrivals and range extensions. On the northern plains, for example, the western kingbird (a flashy gray-and-black flycatcher with a lemon-yellow front) was originally restricted mainly to wooded river valleys in the drainage of the Missouri and the South Saskatchewan rivers. But since the late 1800s, the birds have ventured out of the valleys and, discovering that the world has changed, have expanded through North Dakota and across the plains and parklands of the Canadian prairies. Like land birds crossing an ocean, they have hopped from island to island of wooded habitat, laying claim to a country that in the past had denied them access.

The change that made all the difference for western kingbirds was not so much the expansion of natural islands of brush as the sudden appearance of artificial forests. Ever since the first Arbor Day was celebrated in Nebraska, in 1872, the people of the Great Plains have eagerly bent to the task of cultivating what one prairie arbori-enthusiast referred to as "missionaries of culture and refinement." By which he meant woody plants. Over the decades, the settlers and their descendants have planted literally hundreds of millions of trees and shrubs, including so many shelterbelts that if they were placed end to end, they would reach around the equator and partway back again. Between 1935 and 1958, for instance, a combined total of 32,000 miles (52,000 kilometers) of these rural windbreaks were planted on the grasslands of Canada and the United States, providing a literal hedge against soil erosion. Meanwhile, in both rural and urban settings, the little house on the prairie had become the little house in the woods, as tree-loving people altered the landscape to create habitat for themselves. As these plantings matured, they also provided habitat for western kingbirds and other species of birds, which used them as stepping-stones in their progress across the grasslands.

The list of woodland birds that have extended their breeding range on the prairies includes such formerly rare and now at least locally common species as tree swallows, mourning doves, downy woodpeckers, house wrens, chipping sparrows, black-billed magpies, American crows, American robins, merlins, Mississippi kites, and red-tailed hawks, among many others. The combined effect of these changes is suggested by a recent study of the birds in northeastern Colorado, which showed that almost 90 percent of the species found in the area today did not occur there at the turn of the twentieth century. The vast majority of the newcomers are highly adaptable, go-anywhere woodland species with broad distributions in other parts of the continent and no special affinities to the prairie region.

SEXUAL TENSIONS

ALTHOUGH MOST OF us would not want to do without the robins bobbing on our lawns, the nuthatches hanging from our feeders, or the chipping sparrows chittering in the shrubbery, the abundance of woodland birds on the prairies—and the ecological changes that they represent— are also a subtle menace. What appears at first glance to have been an unequivocal enrichment of life turns out, on closer examination, to pose a covert threat to the specificity, the very prairieness, of the prairie region. For instance, in the old days, before woodland birds could leapfrog across the grasslands at will, the open plains (especially the sparsely treed western rangelands) acted as a barrier between eastern and western species. On one side of the great grassy divide, in the ecotone along the eastern hardwood forests, for example, one would find species such as blue jays, eastern bluebirds, and indigo buntings. On the other side, toward the mountain forests of the west, were closely related birds that occupied much the same ecological niche as their eastern counterparts yet were obviously different: Steller's jays (with black crests rather than blue), mountain bluebirds (with blue breasts rather than red), and lazuli buntings (with rich splotches of cinnamon adorning their plumage). Presumably first separated thousands of years ago, at the time when the Ice Age forests were forced off the Great Plains and the reign of grasses began, these populations had been held apart long enough for each to develop its own genetic specializations. Thus, merely by preventing regular contact, the grasslands had driven the process of evolutionary change and sparked the creation of new species.

No more. These days, with riparian woodlands extending into the prairie heartland along the Platte River, and woodlands dotted hither and yon, eastern and western species are meeting head on. Of a total of thirteen east-west pairs of woodland birds that converge on the Great Plains, seven have been studied in depth and all have been found to interbreed freely along their zone of contact. In some cases, the interchange appears to be stable, suggesting that the two gene pools have been in intermittent contact over the centuries and now coexist peacefully. For example, the yellow-shafted (eastern) and red-shafted (western) phases of the northern flicker meet and merge in the Missouri borderlands of Montana and North Dakota, where they produce orange-shafted offspring just like the ones that John James Audubon saw in the same region in 1843. By contrast, Baltimore and Bullock's orioles have been jostling back and forth across the prairies for the last twenty-odd years, with Baltimores extending 120 miles (200 kilometers) west across the Canadian prairies and up the Platte River, and Bullock's orioles pushing strongly eastward across southern Kansas. Along this fluid zone of interaction, the two species interbreed, producing birds with mix-and-match markings and raising the possibility that the two distinctive populations might eventually merge into one. If this were to happen, the diversity of the world's bird life—the range of answers that life can offer to the questions posed by change—would be incrementally diminished.

Facial markings of the red-shafted flicker, left; *yellow-shafted flicker,* center; *and their hybrid,* right

Meanwhile, another group of prairie birds is already vanishing, and the influx of forest species has likely not been helping. The grassland specialists—ground-nesting birds like meadowlarks and dickcissels, pipits, bobolinks, and short-eared owls—have suffered dramatic population losses during the past forty or fifty years, and probably for longer. (The available data only go back to 1966, when the annual Breeding Bird Survey, or BBS, was initiated.) During that period, the "grassland bird indicator," a composite measure that reflects the populations of two dozen prairie-loving species, has declined by more than 40 percent. In part, this trend is the result of loss of habitat to agriculture and urbanization. But the birds are also declining even where significant fragments of native grasslands still exist, and the encroachment of woody plants is a likely culprit.

In a recent study, researchers working in the cattle country of northwestern Oklahoma compared the advancing tide of woody vegetation (as revealed by a thirty-year sequence of aerial photographs) with the downward trend in grassland birds (as revealed by thirty years of BBS data). They detected a strong, inverse relationship between the two variables. The more trees and shrubs, the fewer grassland birds. But does this relationship hinge on simple appropriation of space, or are other, more devious interactions also at play? Could it be, for example, that the spread of trees has tipped the ecological balance against grassland birds and in favor of their enemies or competitors?

In their search for answers, researchers have pointed the finger of suspicion at one of the prairie's own native-born residents. The brown-headed cowbird, or "bisonbird," as it might better be called, is a small-to-medium-sized member of the blackbird tribe. (Females are drab; males are splendidly iridescent.) In the old days, cowbirds lived with, and even on, the bison, following them wherever they wandered and feeding on the insects that flew up from their hooves. Although food was seldom wanting, this nomadic lifestyle did impose a considerable cost: the birds had no chance to settle down, build nests, and raise young. Instead—remarkably—they acquired the ability to drop their eggs in the nests of other songbirds and fly off, leaving their offspring for somebody else to incubate, feed, and bring up. Accordingly, cowbirds are classified as obligate brood parasites, because they are entirely dependent on other species to perform nursery duties.

The obvious losers in this transaction are the unwitting hosts, which often put so much effort into feeding their greedy guests that they cannot meet the demands of their own young. In one instance, only 12 percent of meadowlark chicks were found to have fledged in parasitized nests, compared with 26 percent in broods with no cowbirds present. Yet up to a certain point, this interference does not pose any real threat, since the hosts have had thousands of years to adapt to nest parasitism. (Even 12 percent of a suitably large clutch may be all that a species needs to keep its numbers up.) But if the rate of parasitism increases, the balance will shift and the host species will slowly but surely begin to slip.

This appears to be exactly what happens, in some instances, when woody plants expand into the grasslands. Each intrusion increases the length of the "shoreline," or edge, along which grass meets woods, the very zone in which opportunists like the cowbird are most successful. The nearer a ground nest lies to a stand of trees, the greater the chance that a cowbird will find it and come calling. This may be either because cowbirds concentrate their search along grassland-woodland edges (where they have the best of two habitats) or because they perch on exposed branches, overlooking the prairies, as they search for places to drop their eggs. In either case, improved success for cowbirds will likely come at the expense of already beleaguered grassland specialists. Thus, the apparently thriving populations of dickcissels on small, brush-invaded prairies in southwestern Missouri are suffering from poor reproductive success, probably because of nest parasitism.

IF YOU PLANT IT, THEY WILL COME

COWBIRDS ARE NOT the only danger that lurks in the shrubby interface between grasslands and woods. These productive ecotones also provide habitat for a variety of mostly small, nosy, nest-raiding mammals, including skunks, weasels, raccoons, foxes, opossums, and Franklin's ground squirrels. Again, the greater the penetration by woody plants, the greater the threat to ground-nesting birds. (Increased predation by small mammals appears to be a major factor in the decline of some grassland specialists.) For the edge-loving mammals, by contrast, the altered face of the prairies, with its patchwork of grass and brush, offers an irresistible wealth of resources. The whole place seems to shout, "Come and get it!" The list of mammalian predators that have responded to this call and expanded their range includes raccoons and opossums (which have been heading north), eastern spotted skunks and gray foxes (slowly pushing west), cougars (heading east along major watercourses), and red foxes (striking out wherever they can in every direction).

It is not only predators that have benefited from the increased interspersion of wooded habitats. A whole range of other mammals—from tree-roosting red bats and tree-

climbing white-footed mice to tree-gnawing porcupines—
have also taken advantage of the shifting tree line. Fox
squirrels (a bright-eyed, bushy-tailed tribe of arboreal rodents
once restricted to open woods along the easternmost fringe of
the plains) have pushed steadily north and west in recent years,
scampering from river valley to farmyard to city street. By
1979, for example, they had established themselves in Regina,
by way of the Souris River. Since then, they have expanded
east up Thunder Creek and north across the Qu'Appelle
River, to within calling distance of the boreal forest.
Meanwhile, their cousins in Nebraska have been advancing
west along the South Platte River to occupy the shelterbelts
and riparian forests of eastern Colorado. If you plant a tree,
you end up with a squirrel.

But perhaps the most remarkable story of all is that of
white-tailed deer. Now a familiar sight almost everywhere
on the prairies, these alert, graceful creatures were originally
native to the hardwood forests of eastern Canada and the
United States, where they made themselves at home in grassy
meadows and open woods. As recently as the mid-1800s, this
was the white-tails' world and they seldom, if ever, ventured
westward onto the Great Plains. Although a small population
seems to have colonized the riparian forest of the Missouri
River—prime habitat, one might think—they were rare even
there and were seldom noticed by early visitors. Instead, the
journals of prairie exploration are filled with references to the
abundant herds of elk that, in those halcyon days, flourished
in the riparian forests and hill country. (Although the prairie
subspecies of elk was decimated in the 1880s, due to hunting and habitat loss, Rocky Mountain elk have since been
introduced to the Black and Cypress hills and several other

refuges. The native subspecies persists in parts of southern and central Manitoba and Saskatchewan.)

It was only when the explorers struck out across the grasslands that they really started to notice deer: a small herd of does at the mouth of a brush-filled gully, a lone buck silhouetted along the crest of a rocky ridge. But even here, these sightings were not of white-tails. These animals were mule deer, a typically western species that ranged—and still ranges—from the woody draws and breaks of the prairies through the Rockies and all the way to the Pacific. Most at home in rough, open country marked with tangles of scrubby brush, the mule deer is characterized by its large, alert ears and hair-trigger nerves. It is perpetually poised to bound, or "stott," with its peculiar, stiff-legged gait, up and over obstacles and out of harm's way.

The white-tailed deer, by contrast, is adapted to a gentler landscape of meadows and leafy groves, where a frightened deer can melt into the shadows. At the first hint of danger, the flaglike tail shoots up, sending a conspicuous, waggling warning to friends and relatives. And away they all run, into the nearest woody cover. (Female white-tail and mule deer both typically live in same-sex groups, often with their mother, sisters, and aunts. Bucks tend to be solitary, except during the rut, and seldom associate with close relatives.) Although of little benefit on open grasslands, this defensive strategy has proven remarkably well suited to the mosaic of croplands, hay fields, and scattered groves that has sprung up on the prairies since settlement. Despite a turn-of-the-century crisis, in which both mule and white-tailed deer were almost exterminated by overhunting, white-tails have come back in abundance and extended their range across the Great Plains and beyond. They are now by far the most numerous wild ungulate in North America, with a continental population that likely stands at four or five times what it was when the *Mayflower* landed.

Meanwhile, the deer of the west—the mule deer—have been placed at a disadvantage by the changing landscape, including the expansion of trees, and have suffered widespread population declines since the 1990s. In Montana, for example, where mule deer traditionally outnumbered white-tails by a ratio of 2:1, the species are now approaching equality. Since muleys are typically able to dominate whitetails (to put them on the run), this shift presumably is not the result of competition for food or other resources. Instead, the declining fortunes of mule deer are signs and symptoms of the loss of the scrubby, open country that the species loves.

CHAPTER 8

THE NATURE OF FARMING

Someday we shall look back
on this dark era of agriculture and shake our heads:
How could we have ever believed
that it was a good idea to grow our food with poisons?

JANE GOODALL, *HARVEST FOR HOPE:*
A GUIDE TO MINDFUL EATING, 2005

T HE FIRST HINT that something strange was about to happen was the sound: a loud metallic hissing, almost a clattering, that seemed to rise out of the very ground. We were out on the bald-headed prairie, scuffing along a dirt track through a heat-stressed field of hay. In front of us lay a farm dugout, a pond that had been gouged out of the earth, its banks a moonscape of heaped dirt. The soil around the crater was gray and thin, bare of vegetation and brittle with drought. With every step, tiles of earth crumbled beneath our boots.

The ominous, high-pitched buzz assaulted us from every side, spinning up from our ankles and ricocheting off the dirt banks. Was this a warning? Should we turn around and head back across the field, to the safety of our vehicle?

It couldn't be rattlesnakes, not unless there were dozens of them, circling around us. But that was too farfetched to be credited. What else? A vague thought stirred in my mind. Hadn't I read something somewhere about "rattlesnake grasshoppers"? Could mere insects really set the air ringing with such menace? Apparently so, for a moment of observation confirmed that the margins of the dugout were jumping with grasshoppers—some with yellow bodies or scarlet wings but most of them so drab that they blended into the gray earth. I could stare straight at them and not even guess they were there. Then, the second I moved a muscle, these armored forms broke cover, flung themselves into the air, and spurted away with a loud clacking of their wings, only to fall back to the ground at some little distance and disappear again. As they landed, they sent another cloud of noisemakers racketing into flight, and so on and on, until the whole place was jangling. These were the creatures that had set our nerves on edge and made us consider turning back.

But before there was time to relax, a sudden burst of noise pierced through the ambient buzz, an intense and aggressive-sounding rattle. I looked up to see what I assumed must be a grasshopper suspended in midair above the top of the bank, its wings blurred with effort. The insect hung there, clattering loudly, ablaze with reflected light, for what seemed to be several long minutes, before dropping back to the ground and out of sight. A few minutes later, another frenetic performer leapt into the sky and enacted the same remarkable feat. Alarm gave way to amazement. What was going on here? What kind of grasshoppers were these and what on earth were they doing?

If you have ever tried to catch a grasshopper on a sun-warmed slope, you will appreciate the absurd scene that followed. Time and again, clumsy human hunter stalked spring-loaded insect—run up close, crouch, and pounce—only to come away empty-handed. But finally, an unwary specimen, preoccupied with munching on a dry stalk of grass, permitted itself to be captured and carried away for examination. A brief check in a guidebook confirmed that this was, in fact, a rattlesnake grasshopper or, as it is perhaps more correctly known, a Carlinian snapper. The insect in hand was a female, recognizable not only by the pronged egg-laying organ, or ovipositor, at the end of the abdomen but also by her quiet, businesslike demeanor in the field. You won't find a female Carlinian snapper making a spectacle of herself. That risky behavior is left primarily to the territorial males, which (as near as anyone can tell) perform their showy display flights to attract potential mates. In so doing, they demonstrate not only their stamina but also their daring, since by drawing attention to themselves, individuals run the risk of being noticed by predators. Tellingly, a small flock of songbirds lined the far bank of the dugout, and the ground was strewn with sleek gray-and-white flight feathers, presumably shed by one or another species of grasshopper-hunting songbirds.

It is worth emphasizing that these life-and-death dramas were being played out not on native

Carlinian snapper

prairie but in an environment that, over the preceding hundred-odd years, had been intensively cultivated. This field was a worst-case situation; land that was totally unsuited for farming—stony, dry, and thin—had been plowed up, eventually abandoned as cropland, and only later planted to crested wheatgrass and reclaimed as a marginally productive hay field. The banks of the dugout, by contrast, had not been reclaimed at all; they were as barren as they had been on the day they were scooped out. Yet in the midst of this human disturbance and disappointment, the wild vigor of the grasslands, as seen in the ancient lineage of the Carlinian snapper grasshoppers, had found room to express itself, in all its strangeness and surprise and fascination.

THE NORTH AMERICAN
AGRICULTURAL REVOLUTION

CROPLAND IS A wild place that has been partially tamed. By definition, it has been stripped of its native vegetation and seeded to one or another of the small, select group of plants that humans have brought under cultivation. (In the entire world, only about seventy species of plants are commonly grown as crops; by comparison, there are five thousand kinds of wild plants on the Great Plains alone.) Agriculture was first introduced to the prairies at least two thousand years ago by immigrants from the hardwood forests of what is now the eastern United States. Sometime around the beginning of the current era, circa 1 CE, these people began tending plots of corn and sometimes beans along the tree-lined river valleys of the tall-grass-prairie region, from western

Missouri to Nebraska. (At one point, around 1000 CE, they had extended their influence west to Colorado and north to the Red River Lowlands of Manitoba.) By crafting an economy that depended not only on small-scale agriculture but on hunting and gathering as well, the people of this Plains Woodland tradition sustained themselves on the grasslands for at least a millennium.

Sometime between 900 CE and 1400 CE, however, disaster apparently struck (perhaps in the form of a severe regional drought) and the Plains Woodland culture appears to have died out. Yet even as this group was declining, another wave of farming-and-hunting people was already on the move. Known as the Plains Gardeners or Plains Village culture, these folks established themselves in scattered clusters of earth lodges along rivers and creeks from the Upper Missouri River south to the Republican River and its tributaries, in present-day Nebraska and Kansas. There they lived by hunting, fishing, gathering, and small-scale agriculture, growing not only beans and corn, as the Plains Woodland people had, but also squash and pumpkins. Known from archeological strata that date back to 1000 CE, this tradition was maintained by peoples such as the Pawnees, Mandans, and Hidatsas for more than eight hundred years. It was only in the mid-1800s, after the introduction of smallpox, that the villages and gardens were abandoned.

Obviously, this pioneering agriculture was responsible for a modest, localized loss of natural habitat. But as minuscule islands of human activity in a vast expanse of native grassland, the gardens did not pose a threat to the functioning of the ecosystem. The natural disturbance regime of the prairies was still operating at full force. Wildfires swept freely across the landscape, as random as lightning and wind, removing dried-out vegetation and making way for fresh green growth to push in. Bison, too, tracked across the plains and generally stirred things up, leaving behind a free-form mosaic of grazed and ungrazed plants, dusty wallows, and splotches of dung. By constantly creating new habitats, these dynamic forces kept the prairie in a continuous state of renewal, able to meet the vital requirements of its many and varied inhabitants.

The introduction of intensive agriculture, however, brought these natural processes to a halt. Within less than a human lifetime, between the 1870s and the 1920s, the bison were killed off, wildfire was brought under control by the settlers, and a mind-boggling expanse of grassland was brought under cultivation. The most hospitable and fertile lands—those with the deepest soils, most generous rainfall, and gentlest slopes—naturally bore the brunt of this agricultural revolution. By the time the dust settled, virtually all of the tall-grass prairie, with its rich tangle of grasses and flowers, had disappeared, replaced primarily by an ever-expanding monoculture of corn. (Originally native to Central America, corn is a warm-season grass, like the native species it displaced, and is suited to the growing conditions in the sunny and well-watered eastern regions of the Great Plains.) In the mixed-grass zone, by contrast, the dominant crop is wheat, a cool-season grass that was first domesticated in the Middle East. Together with oilseeds and a number of specialty crops, such as lentils and chickpeas, wheat now dominates between 15 and 96 percent of the mixed grasslands, the exact percentage depending on local growing conditions. Only land that is utterly unsuited for crops has been left in native grass, a

WHERE HAS ALL THE GRASSLAND GONE?

We often assume that the conversion of grasslands to farmland occurred in the distant and unreachable past. And, of course, it is true that vast expanses of wild prairie were plowed up in the early decades of agricultural settlement. Not surprisingly, the areas with the richest soils and most agreeable weather took the hardest hit. The surviving grasslands, by contrast, are generally ill-suited for growing crops. This is cattle country. If someone advises you to give up beef for the sake of the planet, don't be swayed. Range-fed cattle are the guardians of the prairies.

These surviving vistas of intact grassland hold the future of the Great Plains ecosystem. This is where you'll catch the flash of a bluebird's wing and hear the coyotes' eerie song. This is where the grasslands go about their age-old tasks, sequestering carbon, protecting clean water, and creating habitat for the prairie's unique array of organisms.

And yet, year by year, we continue to lose this precious inheritance, an acre here, a hectare there, to cultivation, energy extraction, and urban development. According to the World Wildlife Fund's Plowprint Report for 2018, the Great Plains have suffered more extensive damage than the Amazon rainforest, having lost a greater proportion of their natural vegetation. In 2017 alone, another 1.7 million acres (nearly 700,000 hectares) of intact grassland were put under the plow, bringing the total destruction in the last decade to 57 million acres (23 million hectares), an area slightly larger than Kansas.

circumstance that becomes more common as one travels to the south and far west. But even there, on the drought-prone ranges of the short grasslands, more than 30 percent of the landscape is under tillage. *See Map 13: Major Crops of the Great Plains.*

The transformation of these broad expanses of wild prairie into farmland has been an abrupt and radical shift. Where the cycle of life was once driven by natural processes, much now depends on the activities of humans. The seasonal regimen of field work—the annual round of tillage, seeding, spraying, and harvesting—creates a new rhythm of disturbance that native species must cope with as best they can. At the same time, the substitution of a handful of cultivated plants for the riot of native grasses, forbs, and shrubs has reduced both the range of food and the diversity of habitats available to wildlife, with predictable consequences. It is no great surprise that cultivated fields support fewer species of native prairie animals and plants than do native grasslands. For instance, according to one study, a small (27-acre, or 11-hectare) remnant of unbroken land in Saskatchewan provided habitat for twice as many species of butterflies as were found on the entire surrounding acreage of cropland. Similarly, research done in Nebraska found thirteen species of birds nesting on untilled land but just two in a nearby cornfield. In other words, 85 percent of the expected grassland bird life had gone missing.

But for species that can adapt to the changed conditions, these human-dominated "agroecosystems" are a new and wide open frontier. In addition to the crops themselves (and what are they but birdseed, after all?), farming has introduced a number of wildlife-friendly features to the landscape of the Great Plains. Shelterbelts now provide habitat for tree-dwelling species such as squirrels and woodpeckers; fences and telephone wires crisscross the country with a network of elevated perches. Skunks den under abandoned buildings, and porcupines winter in rusted-out cars, finding shelter in places where it did not exist before. These examples could be multiplied many times over. Although farmland is not native prairie and does not support the full array of species that were here before, it very clearly has a vitality all its own. Simply by going about their business, prairie farmers have provided living space for wild animals and plants. In return, they enjoy a daily access to the natural world that makes them the envy of many city dwellers.

UNWANTED GUESTS

THE PRESENCE OF wildlife in an agricultural ecosystem is not always welcome. Farming often appears to be a struggle against nature, just one darn losing battle after another. Take the fight against weeds, for example. Weeds are wild plants that seed and sustain themselves mainly, and sometimes entirely, on cultivated land. Because they are able to fit their life cycle into the routines of agricultural work, they can compete with crops for sunlight, water, nutrients, and space and greatly reduce yields. Although a few weedy species, like quack grass, are perennials, which sprout up from fragments of root, most are annuals, which rely on seeds to reproduce. They tend to be fast growing—able to mature between farm operations (for example, before planting or after the harvest is off)—and to produce large crops of big, well-nourished seeds. Fecundity is their main defense against frequent disturbance. If every plant produces scores of viable offspring, at least a few of them are likely to avoid being weeded out and survive to produce their own superabundant generation of seedlings.

Perhaps because the original prairies were composed predominantly of perennials, with few annual plants, it is uncommon for native species to have weedy characteristics. Instead, our common weeds are mostly exotic species that were accidentally introduced during the transition to agriculture. Though the bane of the farmers' existence, their nutrient-rich seeds now provide a reliable, year-round food resource for small mammals and birds, from mourning doves and horned larks in the summer to snow buntings in the winter. Weeds also provide nourishment for a wide variety of insects, especially larval forms. For example, the leaves of Canada thistle (a plant that, despite its name, was introduced into Manitoba from Europe in the 1880s and has since spread across the plains) have become a staple food of painted lady caterpillars, so much so that the adult forms are now sometimes known as thistle butterflies.

Insects that consume weeds may be welcome; not so those that eat crops. Controlling herbivorous insects is the other major challenge facing agriculture. Again, most species that become pests have been introduced from abroad, including such "bad guys" as the Hessian fly, the European corn borer, the English grain aphid, and—a relative newcomer—the Russian wheat aphid. Many native insects, by contrast, have no chance of becoming economic threats, because they are adapted to feed on species or families of plants that are of no commercial importance. To cite one typical example, the female Harris' checkerspot butterfly lays her eggs exclusively on a wild plant called the flat-topped aster, the only food her caterpillars will accept. Nothing else (not even one of the other seventy-odd species of wild asters) will meet their requirements. So unless flat-topped asters someday become a marketable crop, this is one insect—among many—that will never trouble farmers. Similarly, the larvae of the great spangled fritillary and a number of related butterflies feed only on violet leaves and are unlikely to do anything more worrying than nibble on pots of pansies.

Harris' checkerspot butterfly

Nevertheless, many native insects have made the leap from natural prairie to cultivated fields. Among them is a small black-and-yellow-striped wasp (about the size of a large mosquito) that was originally associated with awned wheatgrass, Canada wild rye, and other hollow-stemmed native grasses. The female wasp bored a hole into these stems and laid her eggs inside. After the larvae hatched, they fed within the stem until autumn, when the plant began to dry out and the insects headed toward the ground. A short distance above the base of the plant, they gnawed a notch around the stem, causing it to bend or break and form a stub. Here, beneath a plug of plant matter, the larvae spun a silk-lined nest to protect themselves through a winter spent in diapause, or insect hibernation. In spring, they resumed development and matured into flying adults, which emerged from the now-dead stems to lay their eggs in the new year's growth of grasses.

When wheat was introduced to the Great Plains, the wasps apparently did not take to it right away or at least did not immediately cause noticeable damage. In Manitoba, for example, they continued to rely primarily on their natural hosts until 1906, a decade after intensive wheat farming had begun. That year, however, perhaps because of a failure of the native grasses, flights of stem-boring wasps began to lay their eggs in wheat. Because they are inclined to lay eggs in the species of plant from which they hatched, they quickly formed a lasting attachment to wheat fields. Their success was enhanced by the absence of several of their natural control agents—predatory insects and parasitic wasps—that for some reason did not make the transition from native grasslands to crops.

And so a pest was born. The wheat-stem sawfly, as this insect somewhat misleadingly came to be known, went on to cause massive crop losses on the northern plains in the 1950s and 1960s. Although its importance diminished after solid-stemmed varieties of wheat were developed, it has recently staged a remarkable comeback. In the past, sawflies infested only spring-sown wheat, a crop that grows through the summer and is harvested in the fall, giving the insects time to emerge, lay eggs, hatch, and initiate larval development during the growing season. But they could not survive in fall-sown, or winter, wheat, which matures more rapidly and dries down before the sawflies are ready. Lately, however, evolution appears to have pressed fast-forward and produced a new strain (or perhaps species) of sawfly that emerges a month earlier than before, allowing it to colonize winter wheat as well. And so the process of adaptation continues.

HOPPER HEAVEN

OF ALL THE insects that have benefited from farming, none have been more successful—or more devastating to human interests—than certain species of grasshoppers. During severe infestations, they may consume crops, gardens, shrubbery, and everything else in their path until, as the book of Exodus puts it, "not any green thing" is left. People who lived through the Dirty Thirties may still remember walking through a green slime of swarming insects. No wonder we tend to think of them as intruders. Yet grasshoppers are not intruders; they are native herbivorous insects that, over the last 50 million years or so, have evolved in concert with the natural vegetation of the grasslands.

Although plants may look like passive victims, they are really small green warriors armed to fight off attack. Grasses, for instance, are girded with strands of abrasive silica that make them hard to chew and digest; leaves coated with hairs have much the same effect. Many forbs, including sagebrush and other aromatics, are laced with substances that block digestion or prevent nutrients from being absorbed. To survive, grasshoppers (like other herbivores) have had to come up with ways of overcoming these defenses, whether by developing more powerful mandibles and larger gizzards for grinding up tough plants or by producing antidotes to specific plant toxins. No one species can rise to every challenge.

The result has been the evolution of hundreds of species of grasshoppers—more than four hundred in western Canada and the United States alone—each of which has overcome one or more sets of obstacles and gained the ability to feed on one or more types of vegetation. Some grasshoppers are

EDIBLE INSECTS

Most grasshoppers breed in late summer and die before winter sets in, leaving the next generation to overwinter as eggs in the soil. There are, however, a number of notable exceptions, among them a smallish gray-to-green insect with bold geometric markings that is somewhat confusingly known as the brown-spotted range grasshopper. This species lays its eggs in midsummer, where-upon they either (a) sit there for an entire year and hatch out the following July—the usual pattern on the northern plains, though no one knows why—or (b) get busy and hatch out immediately, as one might expect and as commonly happens in the central and southern states. In either case, winter soon closes in on the young hatchlings, and they are forced to take shelter under leaf litter or buried alive in the earth.

Come early spring—usually sometime in April—the half-grown hoppers emerge from their winter hideaways, resume feeding (on a diet of sedges and grass), and proceed to grow and develop. But no sooner do they become active than they are set upon by a rush of breeding birds that spill across the plains for the nesting season. So many mouths to feed, so many bite-sized grasshoppers to stuff into them. Because these early-season grasshoppers are available in May and June when other insects are scarce, they serve as a critical resource for many species of grassland birds. By the time the summer is over, a single pair of prairie songbirds and their nestlings can consume up to 11 pounds (5 kilograms) of insects, including as many as 149,000 grasshoppers.

*Brown-spotted
range grasshopper*

so specialized that they can eat only a single type of native grass, while others are restricted to one family of plants. The rattling Carlinian snapper, for example, is thought to feed mainly on wheatgrasses, and Turnbull's grasshopper (the farmer's friend) eats only Russian thistle and other related plants. These highly specialized hoppers seldom, if ever, become pests, mainly because they are such fussy eaters.

By contrast, the small number of grasshopper species that do infest croplands tend to be generalists, with wide habitat tolerances and broad food preferences. For example, the clear-winged grasshopper—a major pest of cereal crops across the north-central plains—is naturally adapted to eat many types of wild prairie grasses, including fescues, bromes, and bluegrasses. And what are wheat and barley but highly nutritious kinds of grass? The introduction of agriculture has provided clear-winged grasshoppers with mile upon endless mile of top-quality sustenance, far better than anything available to them under natural conditions. The network of mowed road margins has also benefited them by providing ideal sites for egg laying. (When laying, a female grasshopper bores a hole into the soil with her abdomen, releases a number of cream-colored-to-orange eggs, and then covers them with a protective secretion. This pod is the overwintering stage for all the pest species.)

Another notorious pest species, the two-striped grasshopper, has had an even better run. A real go-anywhere, eat-anything herbivore, it is equipped to feed not only on grasses but also on a wide range of broad-leafed plants, including everything from mustard to sowthistle and from ragweed to lead plant. It can even eat timber milkvetch, a wild legume (found in the mountains and the Black Hills) so poisonous that a few mouthfuls can kill a cow. This hopper is one tough hombre, a kind of six-legged super-villain. Not surprisingly, it consumes all types of crops, including grains, alfalfa, corn, and garden produce, and lays its eggs in a variety of agricultural environments: roadside ditches, weedy summerfallow, stubble fields, and tame pastures. For species like the two-striped grasshopper, farmland is heaven.

Having provisioned these insects so abundantly, farmers find it no easy task to prevent them from taking over. Grasshopper damage can be reduced by tactics such as early seeding, crop rotation, tillage, and the use of trap strips (green strips planted adjacent to a germinating crop, where grasshoppers can be lured and poisoned). When worst comes to worst, however, producers often end up relying on lethal force, by spraying infested regions with insect poisons. The current generation of insecticides is effective, but it is also crude, killing not only pests but also any other insects or spiders that happen to be in the way. One popular preparation, carbofuran, proved to be so toxic to birds—including several dozen California gulls that died after feeding on poisoned hoppers in Saskatchewan—that it was taken off the market in Canada and the United States as a grasshopper-control agent, though it is still used in other countries and for other purposes. And even when these poisons do what they are supposed to do and kill only insects, they still weaken the web of life by wiping out the food supply for insect eaters. (Neonicotinoid insecticides, which are used ubiquitously as seed coatings, dissolve in the soil and leach into wetlands, where they go on killing.) Thus, the battle against insect pests may unintentionally take a toll on snakes, lizards, mice, skunks, bats, quail, small hawks, and songbirds—all of which need insects to feed their young—among many others.

Ultimately, the course of a grasshopper outbreak is beyond human control. Like other insects, grasshoppers are creatures of the weather. As ectothermic, or cold-blooded, organisms, they regulate their body temperature by seeking sun or shade (for example, by basking at the top of a stem or hiding under a leaf) to keep their tissues in the optimal functioning range. Warm, dry summers give grasshoppers an advantage, because they don't have to spend so much time soaking up the sun and can put their energy into feeding and growth. If sunny weather persists into late summer and autumn, when most species lay their eggs, the scene is set for an explosive population increase. Under optimal conditions, a single female can produce as many as 250 eggs. If the following spring brings more fine weather, the new generation of nymphs, or hatchlings, will prosper and reproduce in their turn, until the whole place is hopping with hoppers.

About the only thing that can rein in this runaway population is a good spell of rain. By preventing the insects from maintaining their body temperature, damp conditions hinder their development and growth. At the same time, humidity encourages the spread of deadly grasshopper diseases. Of special note is a fungal pathogen called *Entomophaga grylli,* which, when conditions are right, can decimate a population of grasshoppers almost overnight. A strange malady if ever there was one, *E. grylli* causes its victims to clamber into the top of plants just before they die. (Perhaps the illness makes them feel chilly and so they climb up to bask in the sun.) There they cling, so many small, faded cadavers grasping onto stems in what can only be described as a death grip.

Forced into the service of their nemesis, the grasshoppers apparently serve as aerial capsules in which the fungus can mature and from which it can broadcast its spores. Although this disease can bring a quick end to a grasshopper outbreak, it requires ten days of warm, wet weather to develop full epidemic force and therefore is not reliable enough to be deployed as a biocontrol agent.

BENEFITING THE "BENEFICIALS"

BECAUSE THEY ARE SO responsive to the weather, grasshoppers and many other insects have a built-in tendency to experience cycles of boom and bust. Yet in a state of nature, it must have been relatively uncommon for even the most voracious among them to cut loose and reach plague proportions. The resources they required for survival—including their preferred food plants and suitable egg-laying sites—were often limited and dispersed. They had to struggle to meet their needs or, frequently, die trying. At the same time, they were faced with perils that included not only disease but a large and varied assault force of spiders and predatory and parasitic insects. Although these natural control agents have been pushed to the sidelines by high-tech agriculture, they are still out there in the ecosystem.

Grasshoppers, for example, are beset by insect foes, including both beeflies and blister beetles. Although beeflies really do look like bees, all hairy and striped, they only have one set of wings, a characteristic that identifies them as members of the order Diptera, the flies. (True bees, by contrast, always have two sets, fore and hind.) Blister beetles are typical beetles, though with long, narrow bodies that are often clad in wonderfully iridescent or striped wing casings, or elytra. (And, yes, they do contain chemicals that can cause blistering.) Although totally unrelated to each other, beeflies and blister beetles have adopted similar reproductive strategies. They lay their eggs in the soil close to grasshopper egg-pods, a resource that the gravid females likely locate by smell. When the larvae hatch, they wriggle around in the dirt and attack the egg-pods, consuming grasshopper embryos by the dozen. Although not numerous enough to control a severe outbreak on their own, they do their part in limiting grasshopper populations.

The same is true of certain species of tachinid flies, which look like houseflies festooned with bristles or hairs. All members of this group are parasitic, which is to say that they typically lay their eggs on—or sometimes in—the larvae of other insects. There the eggs develop and pass through various larval stages before bursting out of their unwitting host as adult tachinids. Some of the species that infest grasshoppers are live bearers, hatching their eggs in their own bodies and then laying their larvae on, or in, the early nymphal stages of their victims. Regardless of how the attack is launched, these parasitic invasions are fatal.

The trick to maintaining populations of these natural pest-control agents is to provide food for them. Even predators and parasitoids have to eat. After they emerge as free-flying adults, beeflies, tachinids, and blister beetles all feed on pollen or nectar from their own preferred flowering plants. In other words, a weed-free, wall-to-wall monoculture cannot support them. Instead, they require, at a minimum, strips and patches of land that are left to grow weedy and wild, whether along roadsides, grassed watercourses, or fence lines. A diversity of plants supports a diversity of "beneficial" insects. For instance, the introduction of wild sunflowers has been known to reduce the population of greenflies on a nearby sorghum field, probably by providing nectar for a particular species of greenfly-killing parasitic wasp. And sometimes the linkages are even more circuitous. There is a certain tachinid fly, for example, that parasitizes the European corn borer, a pest that (as its name suggests) tunnels into cornstalks and destroys them. The tachinid is most effective in the presence of giant ragweed, a plant that is often infested by yet another type of stalk-boring insect. Since the tachinid parasitizes this insect as well, it can survive in its alternate host when European corn borers are rare, and so it is ready and waiting to mount an attack when the pest reappears.

It's easy to see how providing food and alternate hosts for beneficial insects would help control pests. But there may well be more to the story than that. When plants are attacked by herbivorous insects, the leaves release volatile compounds into the air, which waft out from their tissues like distress signals. Predatory and parasitic insects can pick up these chemical messages (or smells) and interpret them, perhaps even to the extent of identifying the species of plant-eating insect that is present. If the predator or parasitoid catches the scent of an edible insect or a prospective host, it will home in on the signal and attack, thereby protecting the plant from extensive damage. The more species of plants that are present, the more varied and stimulating their chemical output will be, as if the sos call were beaming out on every available wavelength. Certain parasitoids have proven to be more effective in controlling pest infestations in mixed plots of corn, soybeans, and weeds, with their intermixed signals of distress, than they are in pure stands of corn, which all send out the same message. Diversity multiplies into complexity in natural systems.

A FONDNESS FOR BEETLES AND BEES

OF ALL THE predatory insects that live in croplands, none are more successful—or more important in controlling pests—than the legions of beetles. To date, more than 350,000 species of these armor-winged insects have been identified in various parts of the world, making them the largest and most diverse order of living animals. (Asked what he had learned about the Creator from a lifetime spent studying His works, the Scottish biologist J.B.S. Haldane is said to have quipped, "He seems to have an inordinate fondness for beetles.") By and large, the species found in farmland belong to the family Carabidae, or ground beetles, a group that includes some 40,000 species worldwide. Of these, an astonishing 2,500 species are known to occur in Canada and the United States: that's 2,500 subtly different ground beetles. The most familiar of their number, the sidewalk beetle, is a European species that was accidentally imported by early settlers and that, ever since, has been fast-footing it through lawns and gardens across the continent. Most of the native carabids are variations on the same theme: small, black, shiny, and intensely busy.

Although the majority of carabid species cannot cope with frequent disturbance and are not found on cultivated plots, the number of species that have made the transition to farmland is impressive. For instance, by the simple expedient of setting wide-mouthed jars into the soil and fitting them with funnels instead of lids, a researcher in South Dakota was able to trap 127 kinds of ground beetles on farmers' fields. And while many of these species were rarities, several were exceedingly abundant, with peak, late-summer populations of 4,000 to 6,000 beetles per acre, or 10,000 to 15,000 per hectare. On one extraordinary occasion (when a heavy rain a few days earlier had moistened the soil and created what must have been ideal conditions), one species of tiny carabid topped the charts at an estimated density of 5.6 million individuals per acre (13.8 million per hectare). No matter how you put it, that's a lot of beetles. Yet for the most part, we don't even know they're out there because they hide from the sun by day—in the soil or under litter and leaves—and only come out to hunt in the cool, moist air of evening.

But under cover of darkness, what carnage! Although a few species of ground beetles feed on plants, most are voracious predators that prowl through the dark, searching for insect eggs, larvae, pupae, and soft-bodied adult forms. While some carabids are thought to be specialists, with their sensory radar tuned to specific prey, others appear to be opportunists that eagerly accept any insect food that comes within reach of their mandibles. As a result, they consume untold quantities of insects that might otherwise damage crops, including wireworms, cutworms, caterpillars, maggots, and aphids, among many others. It would not be overstating the case to say that farmers could not attain profitable yields without the unpaid—and generally unheeded—exertions of ground beetles.

The best way to safeguard the diversity and abundance of these essential "farm animals" is to conserve their habitats. Practically speaking, this means maintaining weedy field margins, grassy watercourses, irrigation ditches, roadsides, fence lines, shelterbelts, pastures, wetlands, and other semiwild refuges where the land is subjected to minimal disturbance. Such areas not only provide safe havens during insecticide application—beetles have been seen running toward field edges to flee from toxic sprays—but also serve as sheltered daytime hideaways. (Dense tangles of vegetation offer humidity and shade.) Perhaps most important, these undisturbed patches also provide essential wintering habitat, where beetles can burrow down into the ground and wait for spring's return. The overburden of vegetation provides protection against extreme cold and may also trap an insulating layer of snow, thereby further improving the insects' chances of survival. In one experiment, a raised, grassy berm, or "beetle bank," was found to shelter a hundred overwintering carabids in every square foot (that's 1,100 per square meter).

Retaining strips and patches of undisturbed land is the minimal precondition for sustaining the web of life in intensively farmed regions. Although it is true that these small, untended corners can harbor weeds and insect pests, any problems are generally far outweighed by the benefits. Given that the vast majority of native species are either beneficial or neutral to farming, reserving space for them becomes a simple matter of economic and ecological good judgment. Besides pest control, there is the additional plus of free pollination services. Although crops like wheat and corn are pollinated gratis by the wind, many others—including sunflowers, canola/rapeseed, lentils, alfalfa, and peas—must be pollinated by insects, most notably, by bees. Unlike the colonial honey bee (another import from Europe), most of the thousands of species of native bees are solitary nesters. Each female lays her eggs either in holes in wood or, more often, in cells in undisturbed ground, provisioning each with a dollop of a pollen-based mixture. In producing this "bee bread," the female trails pollen from one flower to the next, thereby ensuring that the plants set seed and produce a marketable yield. Thus, simply by conserving areas of semiwild habitat where bees and other wildlife can live, farm people can expect to profit in every sense of the word.

FARMING AND BIRDS

IF THE INTRODUCTION of intensive agriculture constituted a revolution in the ecology of the Great Plains, prairie farming itself has since undergone radical change. A lot has happened since the horse-and-buggy days. The original patchwork of small, diversified homesteads, with a family on every farm, has gradually been transformed into a sparsely populated landscape of large-scale, specialized operations. Where once there were fields of oats for the draft horses and pastures for the family cow, there are now entire sections devoted exclusively to the production of cash crops. Bigger machinery has called for bigger fields, with clean straightaways and right-angle turns; it's hard to steer something as big as a house around the land's natural contours. Squeezed between high costs for fuel and other necessities and ruinously low prices for what they grow, farmers increasingly rely on government income-support programs. Because their entitlements are typically based on acreage under cultivation, they have been pushed to work every square inch of land and to view natural areas as "idle" and "unproductive."

No one knows, in any detail, how these successive changes in the agricultural landscape have affected populations of beneficial insects on prairie farms (though a sharp decline in many species of native pollinators is cause for alarm). Fortunately, however, researchers have been more successful in tracking the fortunes of farmland birds. The greater prairie chicken, for example, was originally a bird of the tall-grass prairie and the neighboring prairie-and-oak savannas. As agriculture spread, this handsome brown-and-buff-barred grouse expanded its range widely, advancing west to the foothills of the Rockies, south to the Gulf of Mexico, and north to the Aspen Parkland. When agriculture was in its frontier phase, this whole broad sweep of country catered to the chickens' needs, combining grasslands for cover with grain fields for foraging. Soon there were more prairie chickens on the Great Plains than there had ever been before. For example, the *Omaha Republican* reported that, on September 6, 1865, two hunting parties led by Captains Kennedy and Hoagland bagged a total of 708 prairie chickens (together with 6 quail, 10 snipe, 14 hawks, 16 ducks, and 2 rabbits). Such one-day totals were not uncommon.

Although protective laws were quickly invoked, enforcement was lax and the population of prairie chickens inevitably began to pay the price. Yet even after the slaughter ended, sometime around World War 1, prairie-chicken numbers did not rebound. By then, the look of the country had shifted from a seminatural landscape, with small patches of cultivation—perfect prairie-chicken habitat—to an intensively farmed ecosystem with small remnants of grassland. The result has been the slow and painful disappearance of the species from four Canadian provinces and eight American states, along with the fragmentation and contraction of the birds' remaining range. Several key populations continue to diminish.

The point here is not just the plight of the greater prairie chicken, as compelling as that is. It is the ongoing narrative of cause and effect, as farming evolves from one stage to the next. Every modification of agricultural practices produces ecological consequences, destroying habitat for some species and, with luck, creating it for others. Thus, the very changes that spelled doom for prairie chickens were to the advantage of other birds, which flocked in to take advantage of novel habitats, including shelterbelts, road allowances, and field margins. Today, several decades further into the story, these strip habitats in their turn have all but disappeared, as farmers strive to push production to the limits. Since the 1930s, it is estimated that as much as 80 percent of the field margins in the central United States have been brought under the plow, and the same trend can be seen across the Great Plains in general.

The consequences of these changes are not difficult to foresee. A study conducted in the farm country of north-central Iowa found that the region was home to a fabulous diversity of breeding birds, close to a hundred species in total. Of these, almost half nested in woody fencerows and windbreaks, a quarter in wetlands and sloughs, and another half dozen or so in hay fields. Removing any of these elements from the landscape removes its resident birds. By the time only the crops and weedy strip habitats are left, only eighteen species remain. And without the weedy margins, there are just three survivors: the vesper sparrow, the horned lark, and the killdeer. (Because these species are adapted to nest on bare or sparsely vegetated prairie and to breed early in the season, often before field work begins, they can succeed right out in the crops where few other breeding birds venture.)

Horned lark   *Killdeer on nest*

The more the agricultural landscape is simplified, the fewer species of birds it can support. It's not exactly rocket science. Among the common farmland birds that are now declining in parts of the Great Plains are such stalwarts as American crows, European starlings, house sparrows, mourning doves, common grackles, and American goldfinches.

At the same time, happily, some recent trends in agriculture appear to be more positive for our feathered friends. One is the widespread adoption of no-till and minimum-till farming methods. Rather than cultivate the land—leaving the soil bare and exposed—the Gospel of No Till wisely asserts that the ground must always be covered by vegetation. To this end, stubble and other crop residues are left in

FAT OF THE LAND

Until the early 1940s, lesser snow geese made a leisurely journey up the Missouri Valley in spring from wintering grounds on the Gulf Coast to nesting sites in the Arctic, pausing to feed in shallows and marshes along their route. On the return trip in fall, however, most made the journey more-or-less nonstop and were seldom seen along the river. That began to change in the 1950s, when flurries of snow geese began showing up along the Missouri in both seasons.

This change appears to have been caused by a change in farming techniques. Until the 1950s, corn was generally harvested by hand, a method that was laborious but thorough. The newfangled mechanical pickers and combines, by contrast, left the countryside strewn with top-quality goose feed. Thus, the table was set for snow geese (and other waterfowl), which moved out of the wetlands to gorge on the spilled corn.

As irrigation allowed corn production to expand, the geese expanded, too, eventually shifting their principal spring migration westward toward central Nebraska. Each spring, not just flurries but blizzards of snow geese spill into the wetlands of the Rainwater Basin and Platte River Valley, where they build up stores of fat and protein for their journey north.

But the geese may not be faring as well as they used to, due perhaps to more efficient corn harvesting techniques. Concern has also been voiced about the shift from corn to soybeans—poor provender for geese. Since genetically modified soybeans are immune to herbicides, they can be sprayed for millet and other weeds, thereby removing an alternate source of food for migrating geese and other seed eaters.

place after harvest for direct seeding the following spring, and herbicide spraying, instead of tillage, is typically used to control weeds. Designed primarily to prevent soil erosion, this technique also provides nesting cover for certain species of ground-nesting birds, including meadowlarks, pheasants, and grasshopper sparrows. (To return to Iowa again as an example, no-till cornfields in that state have been found to offer nesting sites for a dozen species of birds—up from the expected three—and the nests are packed in at many times the usual density. In this respect, no-till achieves the same benefits as organic farming.)

There is always the worry, however, that these lively no-till fields may ultimately function as ecological traps, first attracting birds in large numbers and then subjecting them to seeding, spraying, and other nest-wrecking disturbances. Just because a species is present doesn't mean that it is managing to reproduce. What's more, if the new soil-holding techniques enable producers to cultivate steep slopes and rough country that could not be farmed in the past, then no-till may be responsible for a net loss of quality habitat.

Fortunately, these ambiguities do not cloud the promise of another tactic: the removal of marginal farmlands from crop production. This process began in earnest during the Dirty Thirties, when many of the dustiest lands in the Dust Bowl were brought under public management as National Grasslands in the United States and as PFRA (Prairie Farm Rehabilitation Administration) community pastures in Canada. Thereafter, progress was sporadic, with periods of reversal, a continuing loss of native grassland, and a gradual extension of the acreage under cultivation. In 1985, however, the cause of biodiversity received an unintended boost when the United States introduced the Conservation Reserve Program (CRP). By contracting with farmers to put marginal farmlands into permanent cover (usually either introduced or native grasses) for a specified number of years, government officials hoped to protect the soil from erosion and simultaneously put some cash into the pockets of farmers who had been hit by the collapse of world markets. As good luck would have it, the program also ended up offering a helping hand to another group of hard-pressed prairie residents, including dabbling ducks, game birds—notably greater prairie chickens—and grassland songbirds. Research in the Dakotas suggests that CRP lands provide nesting habitat for around 5 percent of the states' western meadowlarks and more than a quarter of grasshopper sparrows and sedge wrens.

Alas, the government giveth and the government taketh away, and these gains are not secure. In the United States, support for conservation under the federal Farm Bill has wavered in recent decades, with unreliable budgets and changing priorities. In Canada, by contrast, the federal government discontinued the PFRA pasture program as a cost-cutting measure in 2011, ending a remarkable eighty-year run of national investment in grassland conservation.

CHAPTER 9

LONG-RANGE FORECAST

For a long time it has come back to us in wishes,
this great prairie and these beautiful canyons.
KIOWA ELDER AND U.S. ARMY SERGEANT I-SEE-O,
AMARILLO, TEXAS, 1924

IN A CENTURY when the natural world is slowly dying all around us—when wildness has been pushed to the margins—the wide open spaces of the Great Plains are a landscape of hope. Here is an ecosystem that has experienced the full onslaught of modernization in one brief historical instant and that, though battered and torn, still inspires us with its splendor. This is a country filled with light. It is a place where city streets flow out onto the prairie and draw us along until, almost before we know it, we find ourselves rolling down a dusty gravel road, with warm gusts of meadowlark song blowing in through the open window. It is a land where the seasons surge over us like tides, from the sudden upwelling of spring to the languid heat of summer and from the rushing retreat of autumn to the great sparkling silence of winter.

Look up into the darkness of a prairie night and you will see the universe streaming with stars. Suddenly, it becomes possible to picture yourself on the third planet out from the sun, traveling through the mystery and wonder of whatever is out there.

The prairie opens us to the immensities of space and time. Like few other places on Earth, it reminds us that life operates within broad horizons, with sight lines that extend from the past through the present and into the future. Just as the buffalo prairie is gone, though not forgotten, the countryside that we see before us is even now being transformed into the living landscape of tomorrow. As we look at the world that we have inherited from our ancestors, it is impossible not to think of the generations who will come after us. The wild prairie that we leave to them will be our legacy.

Admittedly, the trends of recent decades have not been encouraging. Although the big plow-down of the settlement era is behind us, native prairie is still being lost year by year and bit by bit, whether to cropland, wind farms, strip malls, or rural subdivisions. According to one recent assessment, about 425,000 square miles (1.1 million square kilometers) of natural grassland have been destroyed in the western United States in the last 150 years. Of these losses, almost 10 percent—42,000 square miles (110,000 square kilometers), an area nearly half the size of Wyoming—were incurred between 1950 and 1990. Although the rate of conversion has slowed in recent decades, the destruction has not stopped. We can rip up ten-thousand-year-old grassland in an instant, but it is beyond our powers to create it.

GOING, GOING...

There are more than 460 "species of concern" on the Great Plains.
These are organisms faring so badly in population and reproductive success that they are feared to be heading toward extinction.
Although compilations vary, the list includes around 18 species
of crayfish and other crustaceans, 19 species each
of reptiles and amphibians, 21 birds (the whooping
crane, for example, and the lesser prairie chicken),
26 mammals (including the black-footed ferret and
the swift fox), 33 fishes (sturgeon, minnows, darters,
shiners, and chubs), 41 snails and mussels, 57 spiders
and insects, and well over 200 kinds of plants (the
blowout penstemon, found only on active sand dunes,
the delicate prairie fringed orchids, and many others).

*Western prairie fringed
orchid, at risk*

   One species that is absent from this disheartening
roster is the black-tailed prairie dog, though it also
deserves mention. After initially declaring the species
endangered back in 2000, the U.S. Fish and Wildlife Service
changed its mind a few years later. The prairie dog is no longer
a candidate for listing. Meanwhile, the animals' plight continues to
affect dozens of other species that are specialized to live in and
around prairie dog towns, including such vulnerable birds as the
mountain plover, ferruginous hawk, and burrowing owl.

   Species that are habitat specialists,
rather than generalists, are at elevated
risk of extinction because they are not
well equipped for change. It is therefore
not surprising that most of the species at
risk, about 70 percent, are endemics. For
this reason, the Edwards Plateau of Texas,
which has more than its share of unique
organisms, is also a zone of endangerment.

*Blowout penstemon,
at risk*

Almost as worrying as the outright disappearance of native prairie is the degradation of what is left, whether through overgrazing, fragmentation by energy development, or the intrusion of invasive plants. Not surprisingly, this incremental damage—death by a thousand cuts—is taking a toll on species that rely on wild grasslands for their survival. Take birds, for example. A study by the North American Commission for Environmental Cooperation (CEC) identified thirty-two species of birds that are "highly dependent" on the Great Plains of Canada and the United States, in the sense that more than half of the population is found in the region at some time of year. This group includes not only permanent, year-round residents, such as prairie chickens, but also fair-weather friends, like Sprague's pipits and marbled godwits, which are present only during the breeding season. In addition, there is a short list of species—sandhill cranes and common mergansers among them—that crowd onto the southern plains during the winter. By analyzing data from the Breeding Bird Survey (a standardized count conducted each year since 1966 in both Canada and the United States), the CEC determined that more than 60 percent of the species that rely on the Great Plains are declining in abundance. In contrast, 23 percent of woodland species and about

Sprague's pipit, at risk

Mountain plover, at risk

28 percent of all bird species in North America are experiencing losses.

Another recent study focused exclusively on birds that depend on dry prairie uplands—Swainson's hawks, killdeers, meadowlarks, bobolinks, long-billed curlews, and the like. Of the three dozen species under consideration, all but five were found to be in decline. What are these "prairie canaries" trying to tell us about the state of our grassland ecosystems? And if prairie birds are suffering, what confidence can we have that the other critters out there (the rodents and creepy-crawlies, say, that inspire less interest) are managing any better?

THE WESTERN ADVANTAGE

BY NOW, YOU may be wondering what happened to the "landscape of hope" that you were promised a few minutes ago. We're lovers in a dangerous time, and preserving the splendor of the living world is a global challenge. As Bruce Cockburn's lyrics remind us, "nothing worth having comes without some kind of fight." Yes, the trends are disheartening. Yes, the risks are real; but so is the potential for renewal. In spite of everything, native grassland still persists across the Great Plains, whether as isolated remnants in the Corn Belt or as vast sweeps of rangeland in the western provinces and states. And even though many of the prairie's special creatures are under severe stress, it is encouraging to remember that every single one of them is still with us. The only known exceptions are the passenger pigeon, a bird that was driven from superabundance into oblivion by market hunting in the nineteenth century, and the Rocky Mountain locust—the scourge of pioneer agriculture—which is thought to have vanished around 1900. A number of important subspecies have also been lost to the past, including the prairie-adapted races of wolves, grizzly bears, and bighorn sheep, each with its own genetic innovations.

Plains wolf, extirpated

Any human-caused extinction is a cause for deep regret. Yet given the intensity of development on the prairies in the last hundred-plus years, this record is better than anyone could have dared to hope for. It is a reminder that the grass-lands and the species they support are adaptable and tough, capable of coping with blizzards, wildfires, and hundred-year droughts. Within limits, they can even cope with us. Yet natural adaptability on its own would not have been enough to bring the full complement of species through a century of crisis. Faced with rapid and violent change, several of the region's unique life-forms would almost certainly have disappeared without the intervention of conservation-minded people. The survival of species like the bison stands as proof that people who love the prairie and take a stand on its behalf are the last, best hope of the Wild West.

GONE TODAY, HERE TOMORROW

THE STORY OF environmental resistance on the Great Plains goes at least back to 1872 and the bloody era of the bison slaughter. In a year when two million bison were killed for their hides—their carcasses left to rot—a man named Samuel Walking Coyote took the exceptional step of rounding up seven orphaned calves and driving them from the Milk River country to his home on the Flathead Reservation in western Montana. There they and their burly descendants soon found their way into the hands of two local ranchers, Charles Allard and Michel Pablo, the latter a reformed buffalo hunter who wanted to make amends for his involvement in the blood-bath. Over the next twenty years, as the last of the wild herds were exterminated, Pablo and Allard devoted themselves to breeding and rearing their captive stock. By 1891, the two men possessed around thirty-five head, making theirs one of the largest herds of bison in existence. (The other survivors included a handful of stragglers in Yellowstone National Park, plus a couple of dozen private herds in Canada and the United States, with a combined population of two hundred to three hundred bison.)

Proving that conservation can be good business, Pablo and Allard managed their herd successfully for the next thirty years, selling hides and mounted heads and supplying live animals to zoos, parks, and other ranchers. Then, in the early 1900s, a change in government regulations opened the Flathead lands to homesteading—an intrusion that Pablo and Allard were helpless to resist—and there was no longer any place for them or their bison. Was the species finally being pushed into oblivion? Faced with this bleak prospect, a group of prominent eastern conservationists formed the American Bison Society, one of the first national environmental groups in North America, in 1905. Its objective was to establish wildlife refuges and, while there was still a chance, stock them with some of the captive bison. In short order, two small blocks of land were acquired, the Wichita Mountains National Wildlife Refuge in Oklahoma and the National Bison Range in Montana, the latter situated on what until then had been Flathead land, a maneuver that continues to be a bone of contention.

With the refuges in place, whether by fair means or foul, it was time to bring on the bison. A deal was struck with Pablo and Allard for the purchase of their herd, but Congress vetoed it as a waste of taxpayers' dollars. The Bison Society then set to work collecting private donations, in nickels and dimes, and ultimately raised enough funds to purchase a few head of breeding stock. Meanwhile, news that the Pablo-Allard herd was for sale sped north to Banff, Alberta, where it reached the ear of park superintendent Howard Douglas. A conservationist who already oversaw a small herd of bison in Rocky Mountain (now Banff) National Park, Douglas persuaded his superiors in the parks service to buy the Pablo-Allard outfit and transport the animals to Elk Island National Park, near Edmonton. Although it took six successive spring roundups to catch the half-wild beasts, just over seven hundred animals were eventually transferred across the border to relative safety.

Elk Island was supposed to be a stopover on the way to the animals' real home, the brand new Buffalo National Park near Wainwright, Alberta. About fifty of the animals managed to evade capture, however, and had to be left behind at Elk Island. The rest of the animals were duly transported to southern Alberta, where some of them were eventually crossbred with cows in an attempt to make them more amenable to domestication. Most of the bison in commercial production today carry cattle genes from this kind of misguided experiment. So, too, do many of the bison in Wood Buffalo National Park in northern Alberta, where the Wainwright herd—by then 6,700 strong—was transferred during World War II. The escapees at Elk Island, by contrast, are thought to breed true to their ancestral stock, as 100-percent-pure-and-unadulterated plains bison. (Other sources of pure plains-bison genetics include the herds in the Henry Mountains in Utah, Sullys Hill National Game Preserve and Wind Cave National Park in the Dakotas, and Grand Teton and Yellowstone national parks in Wyoming.) Of all the bison alive today, less than 1.5 percent are the Real Deal from a genetic point of view, and it is more by good luck than good management that we have any at all.

If there is a moral to this story, it is that prairie conservation is a high-stakes, high-risk undertaking, with a surprise at every bend and a constant call for new strategies and alliances. Plains bison would not have survived the last century without human help, however bumbling and misguided that "help" has sometimes been. The fact that the species exists today is a tribute to a fractious and imperfect collaboration between ranchers and city slickers, individuals and groups, Indigenous and non-Indigenous people, citizens and civil servants, each group following its own bent—and with a little luck thrown in at critical moments. By engaging the problem from different angles, a diverse coalition of people came up with diverse approaches that, over time, have permitted them to confront a series of unexpected challenges. Just as an ecosystem relies on the interplay among species, so conservation depends on the interplay among people, with all their differences.

These days, the big players in bison conservation are beginning to dream about boosting the recovery effort into a whole new dimension. In their mind's eyes, they can already envisage several herds of pure plains bison, each at least a thousand animals strong, roaming over wide vistas of windswept grassland. (Canadian Forces Base Suffield in Alberta and the Charles M. Russell National Wildlife Refuge in Montana are among the locations that have featured in these scenarios.) Ideally, these populations would be free not only of cattle genes but also of cattle diseases, like anthrax and brucellosis, yet would be subject to a full suite of predators—perhaps grizzlies and certainly wolves—and other natural, evolutionary pressures. That's what it will take, the experts tell us, for plains bison to advance from mere survival toward restoration as the living, breathing embodiment of the prairie ecosystem, capable not only of restoring ecological function but also of "inspiring, sustaining and connecting human cultures." Is this just a pipe dream? Or might this hopeful vision be just the spark we need to bring diverse interests together over a fresh pot of joe and get us talking to one another about the future?

THE BUFFALO TREATY

At gatherings held in Browning, Montana, in 2014 and Banff, Alberta, the following year, members of twelve Indigenous groups met to sign the Buffalo Treaty, a document that reaffirms their kinship with buffalo. "Our on-going relationship is so close and so embodied in us," the treaty states, "that BUFFALO is the essence of our holistic eco-cultural life-ways." By signing, the partner nations and tribes pledged "to welcome BUFFALO to once again live among us as CREATOR intended ... [in order] to nurture each other culturally and spiritually."

"It is our collective intention," the text continues, "to recognize BUFFALO as a wild free-ranging animal and as an important part of the ecological system; to provide a safe space and environment across our historic homelands ... so together WE can have our brother, the BUFFALO, lead us in nurturing our land, plants and other animals to once again realize THE BUFFALO WAYS for our future generations."

The signatories to this historic treaty include the Blackfeet, Siksika, Piikani, Stoney Nakoda Chiniki, Bearspaw, Wesley, Samson Cree, and Tsuu T'ina Nations, together with the Blood Tribe, the Assiniboine and Gros Ventre Tribes of Fort Belknap Indian Reservation, the Assiniboine and Sioux Tribes of Fort Peck Indian Reservation, and the Salish and Kootenai Tribes. Mistawasis Nêhiyawak (Mistawasis First Nation) joined the treaty at a separate ceremony in Saskatchewan in 2017.

MEANWHILE, BACK AT THE RANCH...

BISON CARRY THE romance of the grasslands on their broad shoulders, so it is perhaps not surprising that they continue to inspire people to take action on their behalf. But many other grassland-adapted species have also benefited from human intervention. One thinks, for example, of the swift fox, a small, bat-eared, rodent-hunting canid that was once widely distributed across the Great Plains but that, after settlement, became severely diminished in both population and range. Thanks in large part to a captive-breeding program initiated by Beryl and Miles Smeeton of the Cochrane Ecological Institute, the species has been reintroduced to an area in southeastern Alberta and southwestern Saskatchewan from which it had disappeared, and a similar recovery is also underway on the Blackfeet Reservation in Montana. Meanwhile, reintroductions are in progress on Kainai (Blood Tribe) lands in Alberta and at Badlands National Park and the Bad River Ranches in South Dakota. At the same time, a host of agencies and individuals are engaged in the conservation of other organisms, from wild bees to rattlesnakes to rare prairie orchids.

"To keep every cog and wheel," the American environmental philosopher Aldo Leopold once wrote, "is the first precaution of intelligent tinkering." But simply retaining all the pieces of the ecosystem will not be enough in itself; we also need places where the cogs and wheels can be put back together. To ensure their long-term survival, grassland species need wild grasslands—broad expanses of native prairie that, through a natural process of disturbance and renewal, are able to maintain a living mosaic of habitats for a full complement of birds and animals. Despite all the historic losses, large areas of more-or-less natural prairie still exist, especially on the rangelands of the northwestern short- and mixed-grass ecoregions in Alberta, Saskatchewan, Montana, Wyoming, the Dakotas, and Nebraska. To this day, there are at least ten regional landscapes on the northern plains that are dominated, as far as the eye can see, by expanses of native prairie. Several of these areas encompass as much as 4,600 square miles (12,000 square kilometers), larger than either Jasper or Yellowstone national parks, making them of real significance for conservation. *See Map 14: Potential Areas for Large-Landscape Conservation on the Great Plains.*

Unfortunately, virtually none of this landscape has been formally protected by law. In fact, across the Great Plains as a whole only about 1 percent of the countryside has been set aside in parks or conservation reserves, less than in any other biome in North America. *See Map 15: National and State/ Provincial Parks on the Great Plains.*

The surviving expanses of native prairie are hard-working landscapes that today, as in the past, provide the basis for western cattle production. But ranches are not wildlife refuges, and over the years, ranchers have made it clear that varmints like prairie dogs, wolves, and wild bison won't find a warm welcome here. At the same time, ranching has placed a value on both wild prairie as grazing land and on the esthetics of broad horizons. For many ranchers, maintaining large expanses of native pasture in productive condition has been a labor of love, as well as an act of economic self-interest. It is a tribute to their efforts that several significant conservation projects, including Grasslands National Park and Old

Man on His Back Conservation Area in Saskatchewan and the Tallgrass Prairie National Preserves in Oklahoma and Kansas, have been established on lands that were previously managed as commercial ranches.

Ranching not only has the advantage of preventing the prairie from being plowed up. It also helps to keep it from becoming fragmented. Anything that takes a bite out of the natural grassland, whether it be a tame pasture, a pricey ranchette, or an oil field with its network of service roads, breaks up the landscape and subtly alters its ecological function. A study conducted in Colorado showed that the subdivision of ranch land into acreages led to the displacement of grassland birds, such as lark buntings and meadowlarks, and their replacement with robins, magpies, and other common-or-garden species. The pattern was the same for carnivores, with coyotes and bobcats ceding their role as predators to domestic pets. Most worrying of all was the discovery that these cut-up tracts of prairie were much more susceptible than working ranch lands to intrusions by a long list of invasive plants, including introduced species such as smooth brome, crested wheatgrass, leafy spurge, and dozens of others. Given an inch of bare ground, these aggressive weeds will literally take a mile, eventually overwhelming and choking out the natural vegetation. The more lines of disturbance are scratched across the country, the more access these invaders gain and the more quickly the native prairie is forced to give way.

One way to prevent these losses is to keep livestock on the range and to maintain the historical use of the land for grazing. Perhaps the best-known proponent of this approach to conservation is billionaire-businessman-turned-cowpoke

Ted Turner. In an attempt to disprove the adage that the best way to make a small fortune in the livestock industry is to start with a large one, Turner has invested a portion of his wealth in a string of ranches across the Great Plains, with the intention of showing the world that conservation can pay. Now the largest bison producer in the country, with a combined herd on his various holdings of some fifty thousand head, he makes it a policy to stock his ranches below industry standards, both as a means of saving money (no need to provide extra feed during droughts) and as a safeguard against overgrazing. What's more, through the Turner Endangered Species Fund, he and his family are attempting to improve the prospects of prairie dogs, black-footed ferrets, and other species at risk both on the Turner properties and elsewhere. And finally, he has launched a chain of restaurants under the banner of Ted's Montana Grill, where diners can share his taste for prairie conservation.

The livestock industry has been the subject of intense criticism from environmentalists, often for valid reasons. But not all ranches are created equal. Under thoughtful management, working rangelands can be highly productive for wildlife and, at their best, have been known to support more native species, at higher densities, than are found on nearby wildlife refuges. Buying range-fed beef or bison from a conservation-minded producer has a significance that extends far beyond the dinner table. Happily, these prairie-friendly products are readily available in many parts of the Great Plains, whether purchased at the farm gate or through farmers' markets, health food stores, and other alternative outlets. A number of producers are also offering their range-fed meats to a broad public through the internet, under labels such as

COWBOY CONSERVATION

If there were legal protection for endangered societies as well as species, the ranching communities of the western plains would surely top the list. In one struggling town after another, school enrollments are declining, storefronts are boarded up, and the ranchers who meet on coffee row are no longer young. A uniquely Western way of life is at risk of blinking out.

In the past, conservationists and ranchers have often had a testy relationship. But tempers have mellowed with time, and the old adversaries are increasingly able to find common ground in their concern for the future. Ranchers want to earn a living from ranching, and conservationists want them to succeed, as part of a strengthened and diversified nature-based economy.

In a report entitled *New Directions for the Prairie Economy: Connecting Conservation and Rural Development in the Northern Great Plains*, published in 2009, the World Wildlife Fund presented a twelve-point plan for "fostering nature-based economic development in ways that also support the goals of biodiversity conservation." Key recommendations included: expanding ranch-based ecotourism (notably hunting) to "reward landowners who conserve biodiversity"; developing markets for biodiversity-friendly beef and other grassland products; enlarging ecological reserves to encompass at least 10 percent of the northern plains; expanding support for Indigenous peoples' ecological and cultural aspirations; securing adequate funding for existing conservation programs; and offering compensation to ranchers for the ecological services they provide, such as protecting watersheds, preventing soil erosion, and sequestering carbon.

"Organic Prairie," "Native American Natural Foods," and "Wild Idea Buffalo Company."

Two other strategies for conserving native grassland deserve our attention, as well. The first is a type of legal contract known as a conservation easement. Simply put, an easement is a commitment made by a landowner to exempt a specified parcel of land from future development. In return, he or she qualifies to receive compensation from the sponsoring organization, usually an environmental group (like the Nature Conservancy or Ducks Unlimited) or a government agency (like the U.S. Fish and Wildlife Service). Although the terms of the contracts vary, the restrictions typically include a ban on cultivation, subdivision, and the destruction or degradation of wetlands. Sometimes energy development and road building are also prohibited. These limitations apply not only to the current owner but also to anyone who subsequently acquires the land, thereby achieving permanent protection at the cost of a onetime payment. And even though the payouts are modest, they have been welcomed by ranchers, who find themselves chronically caught between high costs of production and low prices at the livestock market.

A second strategy—setting aside sweeping landscapes for wildlife, in parks and ecological reserves—understandably raises the hackles of private landowners. But with fair compensation for those who choose to sell and the promise of a diversified economy, built around ecotourism, to sustain those who remain on the land, there can sometimes be a meeting of minds, a softening of differences, and a merging of agendas.

GROWING CONSERVATION

THE IMPORTANCE OF protecting the surviving horizons of native grassland can scarcely be overstated. As long as there is grass on the land, there is hope for recovery. But protecting "the best of what's left" cannot be our only priority. To borrow again from Aldo Leopold, relegating wild prairie to the western rangelands is like relegating happiness to heaven; one may never get there to enjoy it. Fortunately, there are more immediate pleasures. The regions that have been altered by farming still shimmer with life. Roadsides are heady with wild roses; geese explode out of farmers' fields; pretty little deer mice patter around barns and granaries. Although farmland is not prairie, it nonetheless provides habitat for many species of native animals and plants. Sometimes these farmyard populations are all that is left, remnants of a world that has otherwise vanished.

Take the case of the tall-grass prairie. Apart from a relatively small enclave in the Flint Hills and Osage Plains of Kansas and Oklahoma (where the soil is too rocky to till), this ecoregion has been almost 100 percent converted to crop production. We're talking about the row-after-row-after-row-ness of the Corn Belt. All that remains of the natural vegetation are small, isolated patches that somehow escaped the plow, leaving a tantalizing glimpse of a country alight with butterflies and bright with flowers. Although some of these fragments are smaller than backyard gardens, each one demands attention and care as a unique example of a critically imperiled ecosystem. (Because of local variations in growing conditions and through pure chance, no two patches have exactly the same species in the same proportions.) Through the combined initiatives of national and local organizations, many of these prairie remnants are now protected by law and are intensively managed to prevent the incursion of woody invasives and other takeover artists. Similar rescue efforts have also begun in the mixed-grass ecoregions.

These islands of survival provide critical habitat for many species at risk: the dickcissel, the regal fritillary butterfly, the ornate box turtle, the prairie rattlesnake, and on and on. But, alas, these small, isolated populations remain under constant threat. If some misfortune befalls them—whether through pesticide drift, excess predation, or prolonged drought—there are no neighboring populations to move in and replace them, with the result that a single disastrous season could wipe them out. The only possible solution is, wherever possible, to create blocks of habitat large enough to support viable populations or to provide corridors between the existing fragments. Hence the current lively interest in prairie restoration. Going against the historical trend of plowing prairie up, conservation-minded people have begun to replant it. This has called for innovation in both equipment and techniques and has sparked the development of a native-seed industry. Creating new prairie is tricky, expensive, and, in terms of species richness, never a complete success, but it is an inspiring step in the right direction.

The cause of prairie restoration has found some unexpected advocates, among them the Iowa Department of Transportation. Iowa is farming country taken to the extreme, with only scant vestiges of native prairie. What the state does have, however, is a go-anywhere grid of roads, all of which have vegetated verges. Taken together, these strips add up to about half a million acres (roughly

200,000 hectares) of unproductive land that requires mowing, spraying, and other regular maintenance. In an attempt to reduce costs in the late 1980s, the transportation authorities began to experiment with the use of native plants, on the assumption that they were adapted to local conditions and could look after themselves. Since then, more than 50,000 acres (20,000 hectares) of roadside have been seeded, a little more every year, to either a four-grass mixture—typically big and little bluestem, sideoats grama, and Indian grass—or to a colorful assortment of native grasses and wildflowers.

The results have exceeded all expectations. In addition to controlling expenses, the flower-rich plantings in particular have become slender oases of life, blooming not only with flowers but also with butterflies. In one case, for example, researchers found five times as many butterflies, representing twice as many species, in the high-quality restorations as in comparable grassy or weedy ditches. This success has inspired the Iowa Transportation Commission to pump millions of dollars into the Living Roadway Trust Fund, making the state a leader in what it calls "eco-logical transportation."

THE NEW GREEN REVOLUTION

THE THOUGHT THAT corridors of prairie might one day run along roads and highways across the Great Plains is cheerful to contemplate. But even if it eventually happens, it will not be enough: life cannot be relegated to the margins. In areas devoted to farming, the working landscape is an ecoregion all its own, where there is a great deal more at stake than crop production. Admittedly, cultivated land provides less-than-ideal habitat for many organisms, notably for those that cannot cope with chronic disruption. Yet for species that have been able to make the transition to the agroecosystem, farmland is home, and what happens there is critical to their survival.

The acceptance of conservation as an everyday aspect of farming practice has been halting. On the plus side, thanks to the object lessons of the Dirty Thirties and subsequent periods of drought, the importance of soil conservation is now well established and has been widely translated into both policy and action. On the debit side, however, conservation of wildlife habitat is often a hard sell. Sometimes producers resist as a matter of principle. Having devoted their efforts to maximizing yields to feed a hungry world, they argue that leaving space "idle" for wildlife is a misuse of productive potential. But while this position might have been irresistible in the 1960s, when an exploding human population faced a net shortage of food, it is not so convincing today. In the intervening decades, a complex suite of developments, some of them halfway around the world, have opened up new and more hopeful options for prairie agriculture.

The achievements of the world's farmers since World War II have been stunning. Between 1950 and 1992, for example, world grain production increased by 170 percent, with only a 1 percent expansion of the area under cultivation. Although many people still go hungry, there is still more than enough food available to feed everyone on Earth. This triumph has come about as the result of a no-holds-barred commitment to maximizing production, known colloquially as the Green Revolution. Sparked by an Iowa-born scientist named

Norman Borlaug, who won a Nobel Prize in 1970 for his work, the revolution was based on the development of high-yield varieties of wheat and corn. And yield they certainly do, provided that they are supplied with ample stores of nitrogen, from artificial fertilizer, and abundant water, typically from irrigation.

Unfortunately, the environmental costs of this high-tech fix are becoming inescapably obvious. On the Great Plains, in particular, the damage has included the depletion of aquifers through irrigation, the poisoning of groundwater and wetlands with agricultural chemicals, and the overfertilization of entire river systems. There is now an enormous "dead zone" in the Gulf of Mexico, caused by an influx of nutrients, mostly nitrogen and phosphorus, from artificial fertilizers applied to farmlands in the Missouri/Mississippi drainage. In the summer of 2017—the worst year to date—these lifeless, oxygen-deprived waters sprawled across 8,800 square miles (22,800 square kilometers).

But perhaps the most serious and unexpected downside of the Green Revolution was the collapse of world markets. As the supply of farm commodities outstripped demand in the late 1960s, prices dropped to levels that hadn't been seen since the Great Depression. In response to this economic disaster, many producers rooted out fencerows, shelterbelts, wetlands, and other uncropped lands in an attempt to increase their salable harvest. And as if pressure from the marketplace were not enough, farmers often received extra inducement from government income-support programs, in which entitlements were based on the area under cultivation. The more you plowed up, the more you stood to receive. The result of these combined forces has been an incremental loss of habitat from the farm landscape and a corresponding decline in the abundance and diversity of wildlife.

Plus ça change, plus c'est la même chose. Since about 2006, world grain prices have come out of their decades-long doldrums and begun spiking up, doubling, tripling, and quadrupling to record-high levels. Although partly driven by global consumer demand, the price rise has really taken off because of the market for biofuels. Crops like wheat and corn, once valued as foodstuffs, have now become feedstock for ethanol production. While the biofuel boom has provided welcome relief for farmers (especially the big players in the game), it has done nothing to ease the pressure on prairie landscapes. In fact, the drive toward industrial intensification has itself intensified. In the United States, for example, thousands—perhaps even millions—of acres of marginal farmland that were seeded to grass under the Conservation Reserve Program in the 1980s and 1990s were brought back under the plow, at an untold cost to grassland birds and waterfowl. Meanwhile, carbon that had been sequestered in the soil is being released when the land is worked, putting the lie to biofuel's promise of a low-carbon future.

The grain-based ethanol industry has developed as an instrument of public policy, backed by millions in public funds. Fortunately, public influence is also being exerted in less ambiguous directions. The last decade has seen the creation of innovative programs (typically underfunded but important nonetheless) that pay farmers for managing their private land for the public good by providing services to the environment. Both the Agricultural Policy Framework in Canada and the Farm Bill in the United States provide compensation to farmers who fulfill specific agreements to reduce effluents and emissions, conserve soil fertility, create habitat for wildlife, or provide other benefits to society.

Of course, long before these programs existed, there were already farmers who set high standards for themselves and managed their lands for conservation. They didn't mind taking a few extra minutes to maneuver around marshes and ponds or waiting to cut their hay until after the ground-nesting birds had gone. These days, a growing number of producers are extending this legacy of care by making a day-in, day-out commitment to farming *with*, rather than *on*, the land, using time-tested methods that mimic the natural ecosystem. Rather than viewing soil as a mere substrate for producing yields, the practitioners of "regenerative agriculture" are literally rethinking food production from the ground up. Their first concern is to improve soil health, through nature-approved strategies such as minimum tillage, continuous vegetative cover, rotational grazing, and complex cropping systems. As one "conventional" farmer who has made the switch exclaimed recently, "Great things come from diversity." The benefits are truly impressive: reduced reliance on pesticides and other pollutants, better absorption and retention of moisture, greatly increased biodiversity, mitigation of climate change, and disaster-proofed yields of nutrient-rich foods. Even the relatively simple measure of planting a tenth of a farmed field to strips of prairie plants has been proven to double the diversity of pollinators and birds, while halving the amount of runoff and virtually eliminating soil loss.

MANAGING FOR WILDNESS

ALL OF THESE complexities—of rangeland and farmland, easements and frameworks, opportunities and options—can be summed up in two basic concepts. They are the mantras of prairie conservation. The first is *to protect and enhance wild prairie* wherever it still exists, whether as large, connected landscapes or, where no alternative is left, as one-of-a-kind fragments. The second, often overlapping priority, is *to manage the working landscape for wildness* so that it not only serves the interests of people but also supports a diversity of swimming, flying, walking, and crawling forms of life. Achieving these goals will not be easy. Failing to achieve them will mean a continuing downward trend for many of the prairie's unique ecoregions and species.

These priorities take on even greater urgency in the context of climate change. The grasslands as we know them emerged thousands of years ago, at the end of a three-million-year-long ordeal of glaciation. The difference between the desolation of the Ice Age and the birth of the prairies was a natural warming trend that caused the average global temperature to rise by about 9°F (5°C). Now, we are headed toward a perturbation of similar magnitude, but one that we have caused ourselves. The problem, as we all know, is the thick, insulating blanket of carbon dioxide, nitrous oxide, and other industrial emissions that envelops the Earth and prevents heat from radiating away from the surface. Trapped inside this hotbox, the planet has already experienced a global temperature rise of more than 1°C over preindustrial conditions, and the Great Plains are running ahead of the trend. Data from Environment and Climate Change Canada documents warming on the Canadian prairies of 3.4°F (1.9°C) in less than a human lifetime (1948 to 2016), with more to come. Any reasonably long-lived person born on the northern plains in the 2020s can expect to see the mercury climb by an additional 3.4°F (1.9°C), even if emissions are capped. The increase will ramp up to 11.7°F (6.5°C), if emissions continue to grow unabated. Meanwhile, precipitation is expected to increase in many parts of the region, with less snow but more spring and summer downpours, and more damaging storms. Projections for the southern plains call for cruelly extended heat waves and increasingly severe droughts, interspersed with devastating floods.

The Earth is spiking a fever, and our collective response to date has been to wring our hands. In this crisis, the principles of prairie conservation offer us a way forward, beyond worry and self-doubt. By protecting natural grasslands and rewilding the working landscape, we enhance the ability of Nature to save us from ourselves, as the land regains its capacity to regulate itself. At the same time, we find ourselves on a path toward restoration and health for the prairie and all its creatures, including our own children and grandchildren.

We cannot hold back the future. But we can shape the course of events by engaging—fully, deeply, and passionately—with the present. The survival of the Great Plains Grasslands will depend, in no small part, on our ability to ensure their well-being right now. This is sometimes referred to as the "strategy of no regrets," because the work is worth doing, no matter what happens next. By expressing our love for the prairies in action, we take a stand against despair and hold hands with *all our relations*, now and in the future.

ACKNOWLEDGMENTS

THE AUTHOR IS deeply indebted to the dedicated and creative people who contributed their talents to this project, in particular artist Joan A. Williams, editors Nancy Flight and Paula Ayer, designer Fiona Siu, and cartographers Lorena Patiño and Diane Perrick. This book owes much of its beauty and merit to their efforts.

The information presented here is based on scientific research conducted by biologists at universities, research institutes, and other publicly supported agencies in every province and state across the Great Plains. Although there are too many of them to mention individually, they have each earned a vote of thanks from everyone who loves the grasslands. The experts who served as advisors for the first edition deserve special acknowledgment, since they found time in their busy schedules not only to read all or part of the manuscript but also to offer many invaluable corrections and clarifications. That distinguished panel, identified here by their credentials at the time, includes Richard Cannings, consulting biologist; Sydney Cannings,

NatureServe Yukon; Kenneth F. Higgins, U.S. Geological Survey; John Janovy, University of Nebraska; Dan Johnson, University of Lethbridge; Douglas H. Johnson, Northern Prairie Wildlife Research Center; Henry Murkin, Ducks Unlimited Canada; Paul G. Risser, Oklahoma State System of Higher Education; and Joan A. Williams, consulting biologist. Many others generously shared their time and knowledge, notably: Steve Adair, Ducks Unlimited; M.G. Anderson, Brian Gray, and Rhonda McDougal, Ducks Unlimited Canada; James Basinger and Brian Pratt, Geology Department, University of Saskatchewan; Barrie Bonsal, National Hydrology Research Centre; Jan Bednarski, Benoit Beauchamp, and Arthur Sweet, Geological Survey of Canada; Louis B. Best, Department of Natural Resource Ecology and Management, Iowa State University; Dean Biggins and Mike Lockhart, U.S. Fish and Wildlife Service; Peter Blancher, Bird Studies Canada; Don Buckle, arachnologist, Saskatoon; Bonnie Chasteen, Missouri Department of Conservation Grow Native! Program; Bob Clark and Dean Nurnberg, Canadian Wildlife Service; Meredith Cornett, Minnesota Chapter of the Nature Conservancy; Robert R. Cox, Jr., and Gary L. Krapu, U.S. Geological Survey; Sarah Davies, American Prairie Foundation; Art Davis, Cedric Gillott and Robert Randell, Department of Biology, University of Saskatchewan; Cyndi Evans and colleagues, Prairie State Park, Missouri; Gene Fortney, Nature Conservancy of Canada; Curtis Freese, World Wildlife Fund-U.S.; David A. Gauthier, Canadian Plains Research Center; Rob Gardner, Society of Grassland Naturalists; Robert Gordon, Northern Plains Entomology; Robert Graf and Owen Olfert, Agriculture and Agri-Food Canada; John A. Harrington,

Department of Geography, Kansas State University; Kirk
Henderson, Iowa Native Roadside Vegetation Center; E.H.
Hogg, Canadian Forest Service, Natural Resources Canada;
Sam James, Department of Biology, Maharishi University
of Management; Kris Kendell, Alberta Environment; Anna
Leighton, Saskatchewan Native Plant Society; Steve Malins,
Banff National Park; Richard Manning, author and board
member of the American Prairie Foundation; Mary Ann
McLean, Department of Life Sciences, Indiana State Univer-
sity; Sue Michalsky, range ecologist, Eastend, Saskatchewan;
Russ Miller, general manager of the Turner ranches; Wendell
Morrill, Department of Entomology, Montana State
University; Heather Musgrove, Saskatoon; Dave Naugle and
Thomas M. Power, University of Montana; Kevin Murphy,
Saskatchewan Environment; Mike Phillips, Turner Endan-
gered Species Fund; Keith Roney and Ron Tillie, Royal
Saskatchewan Museum; John Sidle, U.S. National Grasslands;
James C. Trager, Shaw Nature Reserve, Missouri; Elaine
Wheaton, Saskatchewan Research Council; and David and
Lin Zahrt of the Country Homestead Bed and Breakfast in
the Loess Hills of Iowa. I am grateful to you all.

Finally, it is a particular pleasure to acknowledge the day-
in-and-day-out contribution of Keith Bell, whose generosity
and love of life have sustained this endeavor at every step.

FOR MORE INFORMATION

INDIGENOUS SCIENCE AND MYTH

The author gratefully acknowledges the generosity of past generations of Indigenous Elders who allowed some of their knowledge and stories to be recorded in print. The stories included in this book have been edited only for length and are based on the following sources:

page 6: "Buffalo Dance Songs," recorded by Wicita Blain and published by Frances Densmore, "Pawnee Music," *Bulletin of American Ethnology* 93: 1929.

page 37: "The Earth in Unsteady Motion," as told by Little Hill and first published in Henry Schoolcraft, *Information Respecting the Historical Conditions and Prospects of the Indian Tribes of the United States.* New York: J.B. Lippincott, 1852–54.

page 97: Luther Standing Bear, *My People the Sioux.* Lincoln: University of Nebraska Press, [1928] 2006.

page 99: "Into This World of Beauty," Old Lady Horse (Spear Woman) as told to Alice Marriott and published in Alice Marriott and Carol K. Rachlin, *American Indian Mythology.* New York: Cromwell, 1968.

page 139: "The Turtle Continent," as told by an unnamed Lakota person ca. 1910 and published online at chnm.gmu.edu/ exploring/pre_18thcentury/creationstories/assignment.php.

page 177: Black Elk, *Black Elk Speaks: Being the Life Story of a Holy Man of the Oglala Sioux, as told through John G. Neihardt.* New York: Morrow, 1932.

page 241: I-See-O, as quoted by Dan Flores, "Sky, Mesa and the Llano Estacado Bioregion." www-personal.k-state.edu/ ~jsherow/flores.htm.

page 251: "The Buffalo Treaty," from "The Buffalo: A Treaty of Cooperation, Renewal and Restoration," programs.wcs.org/ Portals/175/Documents/The%20Buffalo%20Treaty_2014. pdf?ver=2016-01-29-184835-080.

CONSERVATION CONNECTIONS: PLACES TO BEGIN

A complete catalog of the written sources that were consulted in the preparation of this book would fill a volume of its own. This list is intended as a starting point for further study and engagement. Your bookseller will be able to direct you to references that are appropriate for your area, and conservation groups, including those listed below, can guide you to opportunities to learn and participate.

American Prairie Foundation, americanprairie.org

Great Plains Restoration Council, gprc.org

Intertribal Buffalo Council, itbcbuffalonation.org

Nature Conservancy of Canada, natureconservancy.ca

Southern Plains Land Trust, southernplains.org

The Nature Conservancy, nature.org

World Wildlife Fund, worldwildlife.org/places/northern-great-plains

CHAPTER 1: WHERE IS HERE?

Alt, David, and Donald W. Hyndman. *Roadside Geology of Montana.* Missoula: Mountain Press, 1986.

Boal, Frederick W., and Stephen A. Royle, eds. *North America: A Geographical Mosaic.* London: Arnold, 1999.

Bolen, Eric G. *Ecology of North America.* New York: John Wiley and Sons, 1998.

Bragg, Thomas B. "The Physical Environment of the Great Plains Grasslands." In *The Changing Prairie*, edited by Anthony Joern and Kathleen H. Keeler, 49–81. New York: Oxford University Press, 1995.

Brown, Robert H. *Wyoming: A Geography.* Boulder: Westview Press, 1980.

Fenneman, Nevin M. *Physiography of Eastern United States.* New York: McGraw-Hill, 1931.

———. *Physiography of Western United States.* New York: McGraw-Hill, 1931.

Fung, Kai-lu, ed. *Atlas of Saskatchewan.* Saskatoon: University of Saskatchewan, 1999.

Griffiths, Meland, and Lynnel Rubright. *Colorado: A Geography.* Boulder: Westview, 1983.

Hart, Richard H., and James A. Hart. "Rangelands of the Great Plains before European Settlement." *Rangelands* 19 (1997): 11.

Holliday, Vance T., et al. "The Central Lowlands and Great Plains." In *The Physical Geography of North America*, edited by Antony R. Orme. New York: Oxford University Press, 2002.

Jordan, Terry G., et al. *Texas: A Geography*. Boulder: Westview, 1984.

Kelly, Jordan. "Archaeologists Explore a Rural Field in Kansas, and a Lost City Emerges." *Los Angeles Times,* August 19, 2018. latimes.com/nation/la-na-kansas-lost-city-20180819-htmlstory.html.

McKnight, Tom L. *Regional Geography of the United States and Canada*. Upper Saddle River: Prentice Hall, 1997.

Nelson, H.L. *A Geography of Iowa*. Lincoln: University of Nebraska Press, 1968.

Ostlie, W.R., et al. *The Status of Biodiversity in the Great Plains*. Arlington: The Nature Conservancy, 1997.

Radenbaugh, Todd A., and Patrick Douaud, eds. *Changing Prairie Landscapes*. Regina: Canadian Plains Research Center, 2000.

Rafferty, Milton D. *Missouri: A Geography*. Boulder: Westview, 1983.

Ricketts, Taylor H., et al. *Terrestrial Ecoregions of North America: A Conservation Assessment*. Washington, D.C.: Island Press, 1999.

Samson, Fred B., et al. "Great Plains Ecosystems: Past, Present, and Future." USGS Staff—Published Research, 45 (2004), digitalcommons.unl.edu/usgsstaffpub.

Sims, Phillip L., and Paul G. Risser. "Grasslands." In *North American Terrestrial Vegetation*, edited by Michael G. Barbour and William Dwight Billings, 324–56. Cambridge: Cambridge University Press, 2000.

Trenhaile, Alan S. *Geomorphology: A Canadian Perspective*. New York: Oxford University Press, 1998.

Weaver, J.E., and F.W. Albertson. *Grasslands of the Great Plains: Their Nature and Use.* Lincoln: Johnsen, 1956.

World Resources Institute—PAGE, 2000. "Global Extent of Grasslands." wri.org/publication/pilot-analysis-global-ecosystems-grassland-ecosystems.

CHAPTER 2: DIGGING INTO THE PAST

Anderson, Wayne I. *Geology of Iowa: Over Two Billion Years of Change.* Ames: Iowa State University Press, 1983.

Beaty, Chester B. *The Landscapes of Southern Alberta: A Regional Geomorphology.* Lethbridge: University of Lethbridge, 1975.

Belcher, C.M. "Impacts and Wildfire: An Analysis of the K-T Event." In *Biological Processes Associated with Impact Events*, edited by Charles Cockell et al., 221–43. Berlin: Springer, 2006.

Beveridge, Thomas R. *Geologic Wonders and Curiosities of Missouri.* Rolla: Missouri Division of Geology and Land Survey, 1980.

Bluemle, John P. *The Face of North Dakota: The Geologic Story.* North Dakota Geological Survey Education Series 11, 1991.

Buchanan, Rex, ed. *Kansas Geology: An Introduction to Landscapes, Rocks, Minerals, and Fossils.* Lawrence: University Press of Kansas, 1984.

Chin, Karen, et al. "A King-Sized Theropod Coprolite." *Nature* 393 (1998): 680–82.

Chronic, Halka. *Roadside Geology of Colorado.* Missoula: Mountain Press, 1980.

DePalma, Robert A., et al. "A Seismically Induced Onshore Surge Deposit at the KPg Boundary, North Dakota." *Proceedings of the*

National Academy of Sciences of the United States of America, 2019. doi.org/10.1073/pnas.1817407116.

Dort, Wakefield, Jr., and J. Knox Jones, Jr. *Pleistocene and Recent Environments of the Central Great Plains*. Lawrence: University Press of Kansas, 1972.

Gordon, Anthony. *Geology of Saskatchewan: A Historical Approach*. Regina: Western Extension College Educational Publishers, 1979.

Gould, Stephen Jay. *Wonderful Life: The Burgess Shale and the Nature of History*. New York: W.W. Norton, 1989.

Gries, John Paul. *Roadside Geology of South Dakota*. Missoula: Mountain Press, 1996.

Lageson, David R., and Darwin R. Spearing. *Roadside Geology of Wyoming*. Missoula: Mountain Press, 1988.

Maher, Harmon D. *Roadside Geology of Nebraska*. Missoula: Mountain Press, 2003.

Ojakangas, Richard W., and Charles L. Matsch. *Minnesota's Geology*. Minneapolis: University of Minnesota Press, 1982.

Pielou, E.C. *After the Ice Age: The Return of Life to Glaciated North America*. Chicago: University of Chicago Press, 1991.

Russell, Dale A. *An Odyssey in Time: The Dinosaurs of North America*. Toronto: University of Toronto Press, 1989.

———. *A Vanished World: The Dinosaurs of Western Canada*. Natural History Series No. 4, National Museums of Canada, 1977.

Skwara, T. *Old Bones and Serpent Stones: A Guide to Interpreted Fossil Localities in Canada and the United States. Vol. 2, Western Sites*. Blacksburg: McDonald and Woodward, 1992.

Spearing, Darwin. *Roadside Geology of Texas*. Missoula: Mountain Press, 1979.

Storer, John. *Geological History of Saskatchewan*. Regina: Saskatchewan Museum of Natural History, 1989.

Trimble, Donald E. *The Geologic Story of the Great Plains*. Geological Survey Bulletin 1493, 1980. pubs.er.usgs.gov/publication/b1493.

Troeger, Jack Clayton. *From Rift to Drift: Iowa's Story in Stone*. Ames: Iowa State University Press, 1983.

CHAPTER 3: THE GEOGRAPHY OF GRASS

Axelrod, Daniel I. "Rise of the Grassland Biome, Central North America." *Botanical Review* 51 (1985): 164–201.

Borchert, John R. "The Climate of the Central North American Grassland." *Annals of the Association of American Geographers* 40 (1950): 1–39.

Bryson, Reid A., and F. Kenneth Hare, eds. *Climates of North America*. New York: Elsevier, 1974.

Collins, Scott L., and Linda L. Wallace, eds. *Fire in North American Tallgrass Prairies*. Norman: University of Oklahoma Press, 1990.

Harrington, John A., Jr., and Jay A. Harman. "Climate and Vegetation in Central North America: Natural Patterns and Human Alterations." *Great Plains Quarterly* 11 (Spring 1991): 103–12.

Herring, David. "Dry Times in North America." NASA Earth Observatory, 2000. earthobservatory.nasa.gov/features/NAmerDrought/NAmer_drought.php.

Joern, Anthony, and Kathleen H. Keeler, eds. *The Changing Prairie: North American Grasslands*. New York: Oxford University Press, 1995.

Johnson, S.R., and Aziz Bouzaher, eds. *Conservation of Great Plains Ecosystems: Current Science, Future Options*. Dordrecht: Kluwer Academic Publishers, 1995.

Kunkel, Kenneth E., et al. "Regional Climate Trends and Scenarios for the U.S. National Climate Assessment, Part 4. Climate of the U.S. Great Plains." NOAA Technical Report NESDIS 142-4, 2013.

Laird, Kathleen R., et al. "Greater Drought Intensity and Frequency Before AD 1200 in the Northern Great Plains, USA." *Nature* 384 (1996): 552–54.

McNaughton, S.J., et al. "Interactive Processes in Grassland Eco-systems." In *Grasses and Grasslands*, edited by C. Barnard. New York: St. Martin's Press, 1964.

National Climate Assessment. "Great Plains." 2014. nca2014.globalchange.gov/report/regions/great-plains.

National Oceanic and Atmospheric Administration. "El Niño and La Niña: Frequently Asked Questions." climate.gov/news-features/understanding-climate/el-niño-and-la-niña-frequently-asked-questions.

Nichols, Douglas J., and Kirk R. Johnson. *Plants and the K-T Boundary*. Cambridge: Cambridge University Press, 2008.

Rissler, Paul G. "Grasslands." In *Physiological Ecology of North American Plant Communities*, edited by Brian F. Chabot and Harold A. Mooney. New York: Chapman and Hall, 1985.

Rosenberg, Norman J. "Climate of the Great Plains Region of the United States." *Great Plains Quarterly* 7 (Winter 1986): 22–32.

Sala, O.E., and W.K. Lauenroth. "Small Rainfall Events: An Ecological Role in Semiarid Regions." *Oecologia* 53 (1982): 301–4.

Sandburg, Carl. *Cornhuskers*. New York: Henry Holt, 1918.

Sosebee, Ronald E., et al., eds. *Rangeland Plant Physiology*. Denver: Society for Range Management, 1977.

Tippett, Michael K., and Chiara Lepore. "ENSO and Tornadoes." 2017. climate.gov/news-features/blogs/enso/enso-and-tornadoes.

Trenberth, Kevin E., et al. "Origins of the 1988 North American Drought." *Science* 242 (1998): 1640–45.

———, and Christian J. Guillemot. "Physical Processes Involved in the 1988 Drought and 1993 Floods in North America." *Journal of Climate* 9 (1996): 1288–98.

U.S. Environmental Protection Agency. "Climate Change Indicators. Climate Impacts in the Great Plains." 2017. 19january2017snapshot.epa.gov/climateimpacts/climate-impacts-great-plains_.html.

Woodhouse, Connie A., and Jonathan T. Overpeck. "2000 Years of Drought Variability in the Central United States." *Bulletin of the American Meteorological Society* 70 (1988): 2693–714.

CHAPTER 4: SECRETS OF THE SOIL

Blair, J.M., et al. "Responses of Grassland Soil Invertebrates to Natural and Anthropogenic Disturbances." In *Invertebrates as Webmasters in Ecosystems*, edited by D.C. Coleman and P.F. Hendrix, 43–71. New York: CABI Publishers, 2000.

Buol, S.W., et al. *Soil Genesis and Classification*. 5th ed. Ames: Iowa State University Press, 2003.

Callahan, M.A., Jr., and J.M. Blair. "Influence of Differing Land Management on the Invasion of North American Tall Grass Prairie Soils by European Earthworms." *Pedobiologia* 43 (1999): 507–12.

Capinera, John L., ed. *Integrated Pest Management on Rangeland: A Shortgrass Prairie Perspective*. Boulder: Westview, 1987.

Conant, Richard T. *Challenges and Opportunities for Carbon Sequestration in Grassland Systems: A Technical Report on Grassland Management and Climate Change Mitigation*. Food and Agriculture Organization (FAO), 2010. fao.org/fileadmin/templates/agphome/documents/climate/AGPC_grassland_webversion_19.pdf.

Gregg, Robert E. *The Ants of Colorado*. Boulder: University of Colorado Press, 1963.

Heat-Moon, William Least. *PrairyErth: A Deep Map*. Boston: Houghton Mifflin, 1991.

James, Samuel W. "Soil, Nitrogen, Phosphorus, and Organic Matter Processing by Earthworms in Tallgrass Prairie." *Ecology* 72 (1991): 2101–9.

———, and T.R. Seastedt. "Nitrogen Mineralization by Native and Introduced Earthworms: Effects on Big Bluestem Growth." *Ecology* 6 (1986): 1094–97.

Kohnke, Helmut, and D.P. Franzmeier. *Soil Science Simplified*. Prospect Heights: Waveland, 1995.

Paul, E.A., et al. "Micro-organisms." In *Grassland Ecosystems of the World: Analysis of Grasslands and Their Uses*, edited by R.T. Coupland, 87–96. Cambridge: Cambridge University Press, 1979.

Pavlychenko, T.K. "Root Systems of Certain Forage Crops in Relation to the Management of Agricultural Soils." Ottawa: National Research Council, Canada, No. 1088, 1942.

Reynolds, John W. "Status of Exotic Earthworm Systematics and Biogeography in North America." In *Earthworm Ecology and Biogeography in North America*, edited by P.F. Hendrix, 1–27. Boca Raton: Lewis, 1995.

Rice, Elroy L. "Allelopathy and Grassland Improvement." *The Grasses and Grasslands of Oklahoma*, 90–111. Annals of the Oklahoma Academy of Science, No. 6, 1974.

Stanton, N.L. "The Underground in Grasslands." *Annual Review of Ecology and Systematics* 19 (1988): 573–89. wzw.tum.de/public-html/lattanzi/Lit/Stanton%201988.pdf.

Trager, James C. "An Introduction to Ants (Formicidae) of the Tallgrass Prairie." *Missouri Prairie Journal* 18 (1998): 4–8.

Van der Heijden, Marcel G. A., et al. "The Unseen Majority: Soil Microbes as Drivers of Plant Diversity and Productivity in Terrestrial Ecosystems." *Ecology Letters* 11 (2008): 296–310.

Wheeler, George C., and Jeannette Wheeler. *The Ants of North Dakota*. Grand Forks: University of North Dakota, 1971.

Wolfe, David W. *Tales from the Underground: A Natural History of Subterranean Life*. Cambridge: Perseus, 2001.

CHAPTER 5: HOME ON THE RANGE

Allen, Steve. "Coyotes on the Move: Changes in Distribution of Coyotes and Red Fox in North Dakota." *North Dakota Outdoors* 58 (1996): 6–11.

Brennan, Leonard A., and William P. Kuvlesky, Jr. "North American Grassland Birds: An Unfolding Conservation Crisis?" *Journal of Wildlife Management* 69 (2005): 1–13.

Brown, James H., and Edward J. Heske. "Control of a Desert-Grassland Transition by a Keystone Rodent Guild." *Science* 250 (1990): 1705–7.

Byers, John A. *American Pronghorn: Social Adaptations and the Ghosts of Predators Past.* Chicago: University of Chicago Press, 1997.

Collins, Alan R., et al. "An Economic Analysis of Black-Tailed Prairie Dog Control." *Journal of Range Management* 37 (1984): 358–61.

Copprock, D.L., et al. "Plant-Herbivore Interactions in a North American Mixed-Grass Prairie. II. Responses of Bison to Modification of Vegetation by Prairie Dogs." *Oecologia* 56 (1983): 10–15.

Derner, Justin D., et al. "Are Livestock Weight Gains Affected by Black-Tailed Prairie Dogs?" *Frontiers in Ecology and the Environment* 4 (2006): 459–64.

Dobkin, David S. *Conservation and Management of Neotropical Migrant Landbirds in the Northern Rockies and Great Plains.* Moscow: University of Idaho Press, 1994.

Fleischner, Thomas L. "Ecological Costs of Livestock Grazing in Western North America." *Conservation Biology* 8 (1994): 629–44.

Frank, Douglas A., et al. "The Ecology of the Earth's Grazing Ecosystems." *BioScience* 48 (1998): 513–21.

———, and Samuel J. McNaughton. "Evidence for the Promotion of Aboveground Grassland Production by Native Large

Herbivores in Yellowstone National Park." *Oecologia* 96 (1993): 157–61.

Gillihan, S.W., et al. *Sharing Your Land with Shortgrass Prairie Birds.* Brighton: Rocky Mountain Bird Observatory, 2001.

Goriup, Paul D. *Ecology and Conservation of Grassland Birds.* Cambridge: International Council for Bird Preservation, 1988.

Hanski, Ilkka, and Yves Cambefort. *Dung Beetle Ecology.* Princeton: Princeton University Press, 1991.

Holechek, Jerry L., et al. *Range Management: Principles and Practices.* 5th ed. Upper Saddle River: Prentice Hall, 2004.

Hoogland, John L. *The Black-Tailed Prairie Dog: Social Life of a Burrowing Mammal.* Chicago: University of Chicago Press, 1995.

Johnson, Douglas H., and Alan B. Sargeant. 1977. "Impact of Red Fox Predation on the Sex Ratio of Prairie Mallards." U.S. Fish and Wildlife Service Wildlife Research Report 6. pubs.er.usgs.gov/publication/wrr6.

Jones, Taylor. "Report from the Burrow: Forecast of the Prairie Dog 2017." Wild Earth Guardians. pdf.wildearthguardians.org/site/DocServer/RFTB_2017_final.pdf.

Kamler, Jan F., and Warren B. Ballard. "A Review of Native and Nonnative Red Foxes in North America." *Wildlife Society Bulletin* 30 (2002): 270–79.

Knapp, Alan K., et al., eds. *Grassland Dynamics: Long-Term Ecological Research in Tallgrass Prairie.* New York: Oxford University Press, 1998.

Knopf, Fritz L. "Changing Landscapes and the Cosmopolitism of the Eastern Colorado Avifauna." *Wildlife Society Bulletin* 14 (1986): 132–42.

Koford, Carl B. "Prairie Dogs, Whitefaces, and Blue Grama." *Wildlife Monographs* 3 (1958): 1–80.

Marriot, Alice, and Carol K. Rachlin. *Plains Indian Mythology.* New York: New American Library, 1985.

Minta, Steven E., et al. "Hunting Associations between Badgers (*Taxidea taxus*) and Coyotes (*Canis latrans*)." *Journal of Mammalogy* 73 (1992): 814–20.

North American Bird Conservation Initiative. "State of North America's Birds 2016: Grasslands." stateofthebirds.org/2016/habitats/grasslands/.

North American Bird Conservation Initiative Canada. "The State of Canada's Birds, 2019." Environment and Climate Change Canada, 2019. stateofcanadasbirds.org.

Olson, Kenneth C., et al. "Effect of Grazing Prairie Dog-Colonized Rangeland on Cattle Nutrition and Performance: A Progress Report." *Rangelands* 38 (2016): 29–33.

Polley, H. Wayne, and Scott L. Collins. "Relationships of Vegetation and Environment in Buffalo Wallows." *American Midland Naturalist* 112 (1984): 178–86.

Saskatchewan Wetland Conservation Corporation. *A Land Manager's Guide to Grassland Birds of Saskatchewan.* Regina: SWCC, 2002.

Tannas, Kathy. *Common Plants of the Western Rangelands.* 2 vols. Lethbridge: Lethbridge Community College, 1997.

Vavra, Martin, et al. *Ecological Implications of Livestock Herbivory in the West.* Denver: Society for Range Management, 1994.

Vickery, Peter D., and James R. Herkert, eds. *Ecology and Conservation of Grassland Birds of the Western Hemisphere*. Camarilla: Cooper Ornithological Society, 1999.

Wagner, Frederic H. "Half Century of American Range Ecology and Management: A Retrospective." In *Foundations of Environmental Sustainability: The Coevolution of Science and Policy*, 129–47. Edited by Larry L. Rockwood, et al. Oxford: Oxford University Press, 2008.

CHAPTER 6: WATER OF LIFE

Anderson, Michael G., and Lisa G. Sorenson. "Global Climate Change and Waterfowl: Adaptation in the Face of Uncertainty." *North American Wildlife and Natural Resources Conference 66* (2001): 300–19.

Batzer, Donald P., et al., eds. *Invertebrates in Freshwater Wetlands of North America: Ecology and Management*. New York: John Wiley and Sons, 1999.

Beauchamp, Wendy D., et al. "Long Term Declines in Nest Success of Prairie Ducks." *Journal of Wildlife Management 60* (1996): 247–57.

Bent, Arthur C. *Life Histories of Familiar North American Birds*. birdsbybent.com.

Beyersbergen, Gerard W., et al. "Northern Prairie and Parkland Waterbird Conservation Plan." Denver: Prairie Pothole Joint Venture, 2004.

Bolen, Eric G., et al. "Playa Lakes: Prairie Wetlands of the Southern High Plains." *BioScience 39* (1989): 615–23.

Cross, Frank B., and Randall E. Moss. "Historic Changes in Fish Communities and Aquatic Habitats in Plains Streams of Kansas." In *Community and Evolutionary Ecology of North American Stream Fishes*, edited by William J. Matthews and David C. Heins, 155–77. Norman: University of Oklahoma Press, 1987.

Cummings, K.S., and C.A. Mayer. *Freshwater Mussels of the Midwest. Illinois Natural History Society Survey Manual 5*, 1992.

Dahl, Thomas E. *Status and Trends of Wetlands in the Conterminous United States 1986 to 1997*. Washington, D.C.: U.S. Department of the Interior, Fish and Wildlife Service, 2000.

Ducks Unlimited. "2017 Waterfowl Population Survey." ducks.org/conservation/waterfowl-surveys/2017/duck-numbers.

ECOS Environmental Conservation Online System. "Pallid Sturgeon (*Scaphirhynchus albus*)." ecos.fws.gov/ecp0/profile/speciesProfile?sId=7162.

Fischer, Jesse R., and Craig P. Paukert. "Historical and Current Environmental Influences on an Endemic Great Plains Fish." *American Midland Naturalist* 159 (2008): 364–77.

Fowler, Eric. "Crane Moves: Sandhill Cranes Arriving Earlier, Shifting East on Platte." *Nebraskaland* (March, 2019): 28–35. magazine.outdoornebraska.gov/2019/02/crane-moves/.

Galatowitsch, Susan M., and Arnold G. Van der Valk. *Restoring Prairie Wetlands: An Ecological Approach*. Ames: Iowa State University Press, 1994.

Gilbert, Lee, et al. *Atlas of North American Freshwater Fishes*. Raleigh: North Carolina State Museum of Natural History, 1980.

Johnsgard, Paul A. *Crane Music: A Natural History of American Cranes.* Washington, D.C.: Smithsonian, 1991.

————. *The Platte: Channels in Time.* Lincoln: University of Nebraska Press, 1984.

Johnson, Barry L., et al. "Past, Present, and Future Concepts in Large River Ecology." *BioScience* 45 (1995): 134–41.

McCafferty, W. Patrick. *Aquatic Entomology: The Fishermen's and Ecologists' Illustrated Guide to Insects and Their Relatives.* Boston: Science Books International, 1981.

Murkin, Henry R. *Prairie Wetland Ecology: The Contribution of the Marsh Ecology Research Program.* Ames: Iowa State University Press, 2000.

Pallid Sturgeon Recovery Program. pallidsturgeon.org.

Petrik, Paula. "Remaking the World: A Sioux Creation Story, ca. 1910." chnm.gmu.edu/exploring/pre_18thcentury/creation stories/index.html.

Poiani, Karen A., and W. Carter Johnson. "Global Warming and Prairie Wetlands." *BioScience* 41 (1991): 613–18.

Power, Greg. "The Missouri River System's 'Other' Fish." *North Dakota Outdoors* 56 (1993): 2–9.

Reynolds, Ronald E., et al. "Impact of the Conservation Reserve Program on Duck Recruitment in the U.S. Prairie Pothole Region." *Journal of Wildlife Management* 65 (2001): 765–80.

Richkus, Kenneth D. Northern Pintail Nest Site Selection, Nest Success, Renesting Ecology, and Survival in the Intensively Farmed Prairies of Southern Saskatchewan: An Evaluation of the Ecological Trap Hypothesis. PhD diss., Louisiana State University, 2002.

Ryckman, Fred. "The Confluence." *North Dakota Outdoors* 8 (March 2000): 8–10.

Seburn, Carolyn N.L., et al. "Northern Leopard Frog Dispersal in Relation to Habitat." Amphibians in Decline: Canadian Studies of a Global Problem. *Herpetological Conservation* 1 (1997): 64–72.

Spencer, John R., and Nils Moller Andersen. "Biology of Water Striders: Interactions between Systematics and Ecology." *Annual Review of Entomology* 39 (1994): 101–28.

U.S. Army Corps of Engineers, Omaha District. Missouri River Recovery Program. nwo.usace.army.mil/MRRP/.

Van der Valk, A.G. "The Role of Seed Banks in the Vegetation Dynamics of Prairie Glacial Marshes." *Ecology* 59 (1978): 322–35.

———, and C.B. Davis. "The Seed Banks of Prairie Glacial Marshes." *Canadian Journal of Botany* 54 (1976): 1832–38.

Water Science and Technology Board. *The Missouri River Ecosystem: Exploring the Prospects for Recovery.* Washington, D.C.: National Academy Press, 2002. nap.edu/read/10277/chapter/1.

Watkins, Owen Benjamin. Sustaining the Recovery of Lake Sturgeon (*Acipenser fulvescens*) in the North Saskatchewan River of Alberta. MSc thesis, University of Alberta, 2016.

Weller, Milton W. *Freshwater Marshes: Ecology and Wildlife Management.* Minneapolis: University of Minnesota Press, 1994.

World Wildlife Fund. *Rivers at Risk: The Status of Environmental Flows in Canada,* 2009. assets.wwf.ca/downloads/wwf_canadas_riversatrisk_technicalreport.pdf.

CHAPTER 7: PRAIRIE WOODLANDS

Archer, Steve. "Have Southern Texas Savannas Been Converted to Woodlands in Recent History?" *American Naturalist* 134 (1989): 544–61.

Bird, Ralph D. *Ecology of the Aspen Parkland of Western Canada in Relation to Land Use*. Ottawa: Research Branch, Canada Department of Agriculture, 1961.

Bradley, Cheryl E., and Derald G. Smith. "Plains Cottonwood Recruitment and Survival on a Prairie Meandering River Floodplain, Milk River, Southern Alberta and Northern Montana." *Canadian Journal of Botany* 64 (1989): 1433–42.

Bragg, Thomas B., and Annehara K. Tatschi. "Changes in Flood-Plain Vegetation and Land Use along the Missouri River from 1826 to 1972." *Environmental Management* 1 (1977): 343–48.

Cable, Ted T. "Windbreaks, Wildlife, and Hunters." In *Wildlife and Habitats in Managed Landscapes*. Washington, D.C.: Island Press, 1991.

Campbell, B.D., et al. "A Synthesis of Recent Global Change Research on Pasture and Rangeland Production: Reduced Uncertainties and Their Management Implications." *Agriculture Ecosystems and Environment* 82 (2000): 39–55.

Droze, Wilmon H. *Trees, Prairies, and People: A History of Tree Planting in the Plains States*. Denton: Texas Woman's University, 1977.

Elliott, Phillip F. "Cowbird Parasitism in the Kansas Tallgrass Prairie." *Auk* 95 (1978): 161–67.

Fitch, L., and B.W. Adams. "Can Cows and Fish Co-exist?" *Canadian Journal of Plant Science* 78 (1998): 191–98.

Fox, L. B., et al. Habitat Guidelines for Mule Deer: Great Plains Ecoregion. Mule Deer Working Group, Western Association of Fish and Wildlife Agencies, 2009.

Houston, C. Stuart. "The Spread of the Western Kingbird across the Prairies." *Blue Jay* 37 (1979): 149–57.

———, and Mary I. Houston. "Saskatchewan Bird Species Which Increased with Settlement." *Blue Jay* 55 (1997): 90–97.

Johnsgard, Paul A. "Cranes of the World in 2008: A Supplement to Crane Music." *Papers in Ornithology*, University of Nebraska–Lincoln, 2008.

Johnson, W. Carter. "Woodland Expansion in the Platte River, Nebraska: Patterns and Causes." *Ecological Monographs* 64 (1994): 45–84.

Kauffman, J. Boone, and W.C. Krueger. "Livestock Impacts on Riparian Ecosystems and Streamside Management Implications: A Review." *Journal of Range Management* 37 (1984): 430–38.

Kaul, Robert B., et al. "The Niobrara River Valley: A Postglacial Migration Corridor and Refugium of Forest Plans and Animals in the Grasslands of Central North America." *Botanical Review* 54 (1998): 44–75.

Reily, Peggy W., and W. Carter Johnson. "The Effects of Altered Hydrologic Regime on Tree Growth along the Missouri River in North Dakota." *Canadian Journal of Botany* 60 (1982): 2410–23.

Ripple, William J., and Robert L. Beschta. "Hardwood Tree Decline following Large Carnivore Loss on the Great Plains, USA." *Frontiers in Ecology and the Environment* 5 (2007): 241–46.

Rogers, Paul C., and Darren J. McAvoy. "Mule Deer Impede Pando's Recovery: Implications for Aspen Resilience from a Single-

Genotype Forest." *PLoS ONE*(2018): doi.org/10.1371/journal
.pone.0203619.

Rood, Stewart B., and John M. Mahoney. "Collapse of Riparian Poplar Forests Downstream Dams in Western Prairies: Probable
Causes and Prospects for Mitigation." *Environmental Management*
14 (1990): 451–64.

Streng, D.R., and P.A. Harcombe. "Why Don't East Texas Savannas
Grow Up to Forest?" *American Midland Naturalist* 108 (1982):
287–94.

Svejcar, Tony. "Riparian Zones: 1) What Are They and How Do
They Work?" *Rangelands* 19 (1997): 4–7.

———. "Riparian Zones: 2) History and Human Impacts." *Rangelands* 19 (1997): 8–12.

Wells, Philip V. "Scarp Woodlands, Transported Grassland Soils,
and Concept of Grassland Climate in the Great Plains Region."
Science 148 (1965): 246–49.

Wilson, Roger E. "Succession in Stands of *Populus deltoides* along
the Missouri River in Southeastern South Dakota." *American
Midland Naturalist* 83 (1970): 331–42.

Winter, Maiken. "Nesting Biology of Dickcissels and Henslow's
Sparrows in Southwestern Missouri Prairie Fragments." *Wilson
Bulletin* 111 (1999): 515–27.

———, and John Faaborg. "Patterns of Area Sensitivity in Grassland-
Nesting Birds." *Conservation Biology* 13: 1424–36.

CHAPTER 8: THE NATURE OF FARMING

Altieri, Miguel A. "How Best Can We Use Biodiversity in Agroecosystems?" *Outlook on Agriculture* 20 (2001): 15–23.

Baker, Herbert G. "The Evolution of Weeds." *Annual Review of Ecology and Systematics* 5 (1974): 1–24.

Berry, Wendell. "Hope." In *Fatal Harvest: The Tragedy of Industrial Agriculture*, edited by Andrew Kimbrell, 373–75. Washington, D.C.: Island Press, 2002.

Best, Louis B. "Conservation Tillage: Ecological Traps for Nesting Birds?" *Wildlife Society Bulletin* 14 (1986): 308–17.

———, Kathryn Freemark, et al. "A Review and Synthesis of Habitat Use by Breeding Birds in Agricultural Landscapes of Iowa." *American Midland Naturalist* 134 (1995): 1–29.

———, Robert C. Whitmore, and Gary M. Booth. "Use of Cornfields by Birds during the Breeding Season: The Importance of Edge Habitat." *American Midland Naturalist* 123 (1990): 84–99.

———, Timothy Bergin, and Kathryn E. Freemark. "Influence of Landscape Composition on Bird Use of Rowcrop Fields." *Journal of Wildlife Management* 65 (2001): 442–49.

Bryan, Georgia G., and Louis B. Best. "Bird Abundance and Species Richness in Grassed Waterways in Iowa Rowcrop Fields." *American Midland Naturalist* 126 (1991): 90–102.

Buchmann, Stephen L., and Gary Paul Nabham. *The Forgotten Pollinators*. Washington, D.C.: Island Press, 1996.

Calpas, James, and Dan Johnson. "Grasshopper Management." Agri-Facts. Alberta Agriculture, Food and Rural Development, March 2003.

Camp, Martha, and Louis B. Best. "Bird Abundance and Nesting in CRP Fields and Cropland in the Midwest: A Regional Approach." *Wildlife Society Bulletin* 25 (1997): 864–77.

Carcamo, Hector A., et al. "Farming and Ground Beetles: Effects of Agronomic Practice on Populations and Community Structure." *Canadian Entomologist* 127 (1995): 123–40.

Clark, Robert G. et al. "Making a Living on the Edge: Beneficial Effects of Field Boundary Habitats for Biodiversity and Agriculture." *Canadian Entomologist* 130 (1998): 321–36.

Cox, Robert R., Jr. "Spring-Staging Ecology of Waterfowl in Nebraska—Then versus Now." *Waterfowl* 12 (2000): 18–19. digitalcommons.unl.edu/cgi/viewcontent.cgi?article=1015&context=usgsnpwrc.

Curry, Philip S. *Bumble Bees of Saskatchewan (*Hymenoptera: Apidae*).* Natural History Contributions No. 5, Saskatchewan Museum of Natural History, 1984.

Erwin, Terry L., et al. *Carabid Beetles: Their Evolution, Natural History, and Classification.* London: Dr. W. Junk B.V. Publishers, 1979.

Jackson, Dana L., and Laura L. Jackson, eds. *The Farm as Natural Habitat: Reconnecting Food Systems with Ecosystems.* Washington, D.C.: Island Press, 2002.

Jackson, Wes. "Natural Systems Agriculture: A Truly Radical Alternative." *Agriculture Ecosystems and Environment* 88 (2002): 111–17.

———. *New Roots for Agriculture.* San Francisco: Friends of the Earth, 1980.

Johnson, Dan. "Band-Winged Grasshoppers of the Canadian Prairies and Northern Great Plains." *Arthropods of Canadian Grasslands* 7 (2001): 5–12.

———. "Slant-Faced Grasshoppers of the Canadian Prairies and Northern Great Plains." *Arthropods of Canadian Grasslands* 9 (2003): 5–16.

———. "Spur-Throated Grasshoppers of the Canadian Prairies and Northern Great Plains." *Arthropods of Canadian Grasslands* 8 (2002): 16–25.

Johnson, Douglas H., and Lawrence D. Igl. "Contributions of the Conservation Reserve Program to Populations of Breeding Birds in North Dakota." *Wilson Bulletin* 107 (1995): 709–18.

Kirk, David A., et al. "Past and Current Attempts to Evaluate the Role of Birds as Predators of Insect Pests in Temperate Agriculture." *Current Ornithology* 13 (1996): 175–263.

Kirk, Vernon M. "Ground Beetles in Cropland in South Dakota." *Annals of the Entomological Society of America* 64 (1971): 238–41.

Matson, P.A., et al. "Agricultural Intensification and Ecosystem Properties." *Science* 277 (1997): 504–9.

McCoy, Timothy D., et al. "Conservation Reserve Program: Source or Sink Habitat for Grassland Birds in Missouri?" *Journal of Wildlife Management* 63 (1999): 530–38.

McMaster, D. Glen, and Stephen K. Davis. "An Evaluation of Canada's Permanent Cover Program: Habitat for Grassland Birds?" *Journal of Field Ornithology* 723 (2001): 195–210.

Owens, R.A., and M.T. Myres. "Effects of Agriculture upon Populations of Native Passerine Birds of an Alberta Fescue Grassland." *Canadian Journal of Zoology* 51 (1971): 697–713.

Samson, F.B., and F.L. Knopf. "Prairie Conservation in North America." *BioScience* 44 (1994): 418–21.

Shutler, Dave, et al. "Bird Communities of Prairie Uplands and Wetlands in Relation to Farming Practices in Saskatchewan." *Conservation Biology* 14 (2000): 1441–51.

Soule, Judith D., and Jon K. Piper. *Farming in Nature's Image: An Eco-logical Approach to Agriculture.* Washington, D.C.: Island Press, 1992.

Tilman, David. "The Greening of the Green Revolution." *Nature* 396 (1998): 211–12.

Welsh, R. *The Economics of Organic Grain and Soybean Production in the Midwestern United States.* Henry A. Wallace Institute Policy Studies Report No. 13, May 1999.

Wooley, James B., Jr., et al. "Impacts of No-Till Row Cropping on Upland Wildlife." *North American Wildlife and Nature Resources Conference* 50 (1985): 157–66.

World Wildlife Fund. "The Plowprint Report." worldwildlife.org/projects/plowprint-report.

CHAPTER 9: LONG-RANGE FORECAST

Askins, Robert A., et al. "Conservation of Grassland Birds in North America: Understanding Ecological Processes in Different Regions." *Ornithological Monographs* 64 (2007). doi.10.2307/40166905.

Bagley, Catherine, et al. "Organic Agriculture and the Global Food Supply." *Renewable Agriculture and Food Systems* 22 (2006): 86–108.

Blancher, Peter. *Importance of North America's Grasslands to Birds.* Commission for Environmental Cooperation, 2003.

Brower, Jennifer. *Lost Tracks: Buffalo National Park, 1909–1939.* Edmonton: Athabasca University Press, 2008.

Comer, Patrick J., et al. "Continent-Scale Landscape Conservation Design for Temperate Grasslands of the Great Plains and Chihuahuan Desert." *Natural Areas Journal* 38 (2018): 196–211.

Conner, Richard, et al. *United States Grasslands and Related Resources: An Economic and Biological Trends Assessment.* Prepared for the National Cattlemen's Beef Association, the Nature Conservancy and Ducks Unlimited, 2001. nri.tamu.edu/publications/research-reports/2001/united-states-grasslands-and-related-resources-an-economic-and-biological-trends-assessment/.

Environment and Climate Change Canada. *Canada's Changing Climate Report, 2019.* changingclimate.ca/CCCR2019/.

Freese, Curtis, et al. *New Directions for the Prairie Economy: Connecting Conservation and Rural Development in the Northern Great Plains.* World Wildlife Fund, 2009. bcgrasslands.org/wp-content/uploads/2018/02/newdirectionsprairieeconomy.pdf.

Gates, C. Cormack, et al. *American Bison: Status Survey and Conservation Guidelines 2010,* International Union for Conservation of Nature, Species Survival Commission. cmsdata.iucn.org/downloads/american_bison_report.pdf.

Kimbrell, Andrew, ed. *Fatal Harvest: The Tragedy of Industrial Agriculture.* Washington, D.C.: Island Press, 2002.

Knight, Richard L., et al. *Ranching West of the 100th Meridian: Culture, Ecology, and Economics.* Washington, D.C.: Island Press, 2002.

Knopf, Fritz L., and Fred B. Samson, eds. *Ecology and Conservation of Great Plains Vertebrates.* New York: Springer, 1996.

Lockwood, Jeffrey A., and Larry D. DeBrey. "A Solution for the Sudden and Unexplained Extinction of the Rocky Mountain Grasshopper *(Orthoptera: Acrididae)." Environmental Entomology* 19 (1990): 1194–1205.

Martin, Leanne M., et al. "An Assessment of Grassland Restoration Success Using Species Diversity Components." *Journal of Applied Ecology* 42 (2005): 327–36.

McCracken, Jon D. "Where the Bobolinks Roam: The Plight of North America's Grassland Birds." *Biodiversity* 6 (2005): 20–29. bsc-eoc.org/download/GrasslandBirdsMcCracken .pdf.

Millet, Bruce, et al. "Climate trends of the North American Prairie Pothole Region 1906–2000." *Climatic Change* 93 (2009): 243–47.

Moehrenschlager, Axel, and Cynthis Moehrenschlager. "Population Census of Reintroduced Swift Foxes in Canada and Northern Montana 2005/2006." Calgary: Centre for Conservation Research, Calgary Zoo, 2006.

National Oceanic and Atmospheric Administration. "Gulf of Mexico 'Dead Zone' Is the Largest Ever Measured." 2017. noaa.gov/media-release/gulf-of-mexico-dead-zone-is-largest-ever-measured.

Nelson, Amy. "Bringing Back the Buffalo." Biohabitats: Lessons from Indigenous Traditions and Innovation, 2018. biohabitats .com/newsletter/ecology-culture-and-economy-lessons-from-indigenous-traditions-and-innovation/bringing-back-the-buffalo/.

Northern Plains Conservation Network. *Oceans of Grass: A Conservation Assessment for the Northern Great Plains*, 2004. protectedareas.info/upload/document/ecoregionplan-northerngreatplainconservationassessmentsummary.pdf.

Ostlie, W.R., et al. *The Status of Biodiversity in the Great Plains*. Arlington: The Nature Conservancy, 1997.

Packard, Stephen, and Cornelia F. Mutel. *The Tallgrass Restoration Handbook: For Prairies, Savannas and Woodlands.* Washington, D.C.: Island Press, 1997.

Power, Thomas Michael, and Richard Barrett. *Post-Cowboy Economics: Pay and Prosperity in the New American West.* Washington, D.C.: Island Press, 2001.

Regeneration International. "Why Regenerative Agriculture?" regenerationinternational.org/why-regenerative-agriculture/.

Ries, Leslie, et al. "Conservation Value of Roadside Prairie Restoration to Butterfly Communities." *Conservation Biology* 15 (2001): 401–11.

Samson, Fred B., and Fritz L. Knopf, eds. *Prairie Conservation: Preserving North America's Most Endangered Ecosystem.* Washington, D.C.: Island Press, 1996.

Sayre, Nathan F. *Working Wilderness: The Malpai Borderlands Group and the Future of the Western Range.* Tucson: Rio Nuevo, 2005.

"Science Based Trials of Rowcrops Integrated with Prairie Strips." Iowa State University, 2019. nrem.iastate.edu/research/STRIPS.

Shirley, Shirley. *Restoring the Tallgrass Prairie.* Iowa City: University of Iowa Press, 1994.

Smith, Elwin G., et al. "Profitability and Risk of Organic Production Systems in the Northern Great Plains." *Renewable Agriculture and Food Systems* 19 (2004): 152–58.

Sundarm, Jomo Kwame. "The World Produces Enough Food to Feed Everyone. So Why Do People Go Hungry?" Global Agenda/Future of Economic Progress/Future of Food. World

Economic Forum, 2016. weforum.org/agenda/2016/07/
the-world-produces-enough-food-to-feed-everyone-so-why-do-
people-go-hungry.

Terra Genesis International. "Regenerative Agriculture."
regenerativeagriculturedefinition.com.

Toensmeier, Eric. *The Carbon Farming Solution: A Global Toolkit of
Perennial Crops and Regenerative Agricultural Practices for Climate
Change Mitigation and Food Security.* White River Junction, VT:
Chelsea Green Publishing, 2016.

U.S. Global Change Research Program. "U.S. National Assessment
of the Potential Consequences of Climate Variability and
Change, Mega Region: Great Plains." usgcrp.gov/usgcrp/nacc/
greatplains.htm/.

Williams Dave, et al. "Reseeding Wildflowers to Diversify Grasslands for Pollinators: Guidance for the Great Plains and Midwest
Regions." Portland: Xerxes Society for Insect Conservation,
2018. xerces.org/wp-content/uploads/2018/10/18-021_01_
Interseeding-Wildflowers-Grasslands-Guide_web-1.pdf.

INDEX

Note: Numbered maps can be found in the middle of the book.

DAVID
SUZUKI
INSTITUTE

THE DAVID SUZUKI INSTITUTE is a non-profit organization founded in 2010 to stimulate debate and action on environmental issues. The Institute and the David Suzuki Foundation both work to advance awareness of environmental issues important to all Canadians.

We invite you to support the activities of the Institute. For more information please contact us at:

David Suzuki Institute
219 – 2211 West 4th Avenue
Vancouver, BC, Canada V6K 4S2
info@davidsuzukiinstitute.org
604-742-2899
www.davidsuzukiinstitute.org

Cheques can be made payable to The David Suzuki Institute.